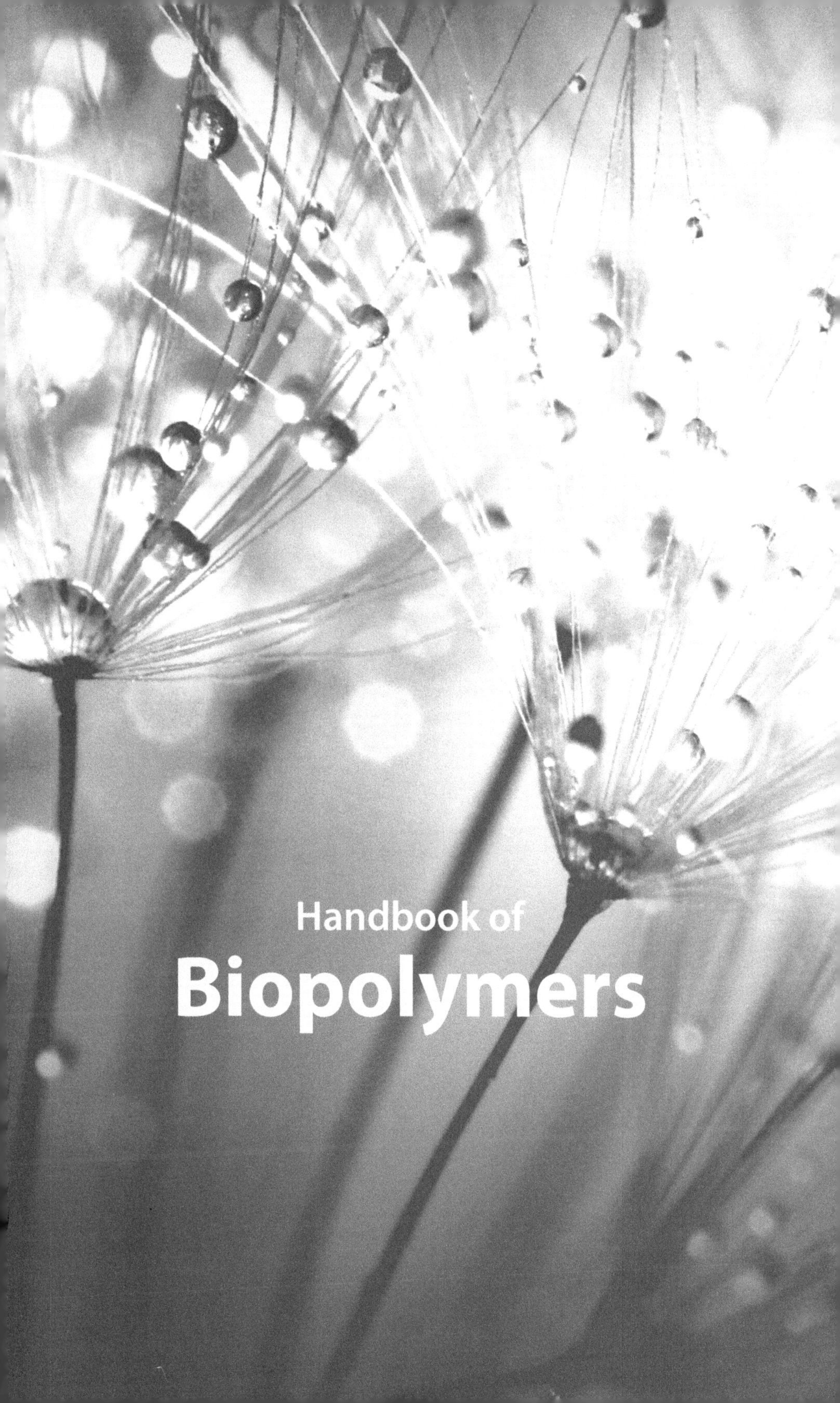

Handbook of Biopolymers

Handbook of

Biopolymers

Advances and Multifaceted Applications

edited by

Shakeel Ahmed
Suvardhan Kanchi
Gopalakrishnan Kumar

PAN STANFORD PUBLISHING

Published by

Pan Stanford Publishing Pte. Ltd.
Penthouse Level, Suntec Tower 3
8 Temasek Boulevard
Singapore 038988

Email: editorial@panstanford.com
Web: www.panstanford.com

British Library Cataloguing-in-Publication Data
A catalogue record for this book is available from the British Library.

Handbook of Biopolymers: Advances and Multifaceted Applications

ISBN 978-981-4800-17-4 (Hardcover)
ISBN 978-0-429-02475-7 (eBook)

Contents

Preface

Biopolymers are the polymers synthesized by living organisms. They can be polynucleotides (such as the nucleic acids DNA and RNA), polypeptides (that is, proteins), or polysaccharides (that is, polymeric carbohydrates). These consist of long chains made of repeating, covalently bonded units, such as nucleotides, amino acids, or monosaccharides. Cellulose is the most common organic compound and biopolymer on the earth. About 33% of all plant matter is cellulose. The cellulose content of cotton is 90%, and for wood it is 50%. Biopolymers can be sustainable and carbon neutral and are always renewable, because they are made from plant materials, which can be grown indefinitely. These plant materials come from agricultural non-food crops. Therefore, the use of biopolymers would create a sustainable industry. In contrast, the feedstocks for polymers derived from petrochemicals will eventually deplete. In addition, biopolymers have the potential to cut carbon emissions and reduce CO_2 quantities in the atmosphere; this is because the CO_2 released when they degrade can be reabsorbed by crops grown to replace them: This makes them close to carbon neutral. Biopolymers are biodegradable, and some are also compostable. Some biopolymers are biodegradable: They are broken down into CO_2 and water by microorganisms. Some of these biodegradable biopolymers are compostable: They can be put into an industrial composting process and will break down by 90% within six months. Biopolymers that do this can be marked with a "compostable" symbol, under European Standard EN 13432 (2000). The packaging marked with this symbol can be put into industrial composting processes and will break down within six months or earlier. An example of a compostable polymer is PLA film under 20 μm thick: Films that are thicker than that do not qualify as compostable, even though they are biodegradable. In Europe, there is a home composting standard and associated logo that enables consumers to identify and dispose of packaging in their compost heap.

This book is about bio-based and biodegradable polymers. It covers a fairly broad range of biopolymers with a strong focus on their applications, simply because of the large global consumption and impact of the latter on the environmental and different life forms on the earth. No matter where in the world plastics objects are thrown out, eventually they find their way to the oceans and continents around the globe. There are many sustainability issues that have been driving the development of monomers and biodegradable polymers from renewable plant resources. Some of those issues are better known to the public, including the cost of the traditional raw material source petroleum, global warming, and environmental damage. A less understood problem is the extent of post-use pollution caused by plastic objects. Of all plastic, metal, and paper collected for recycling, only an estimated 25% actually makes its way to reuse.

There are numerous books about biopolymers covering the scientific research that is enabling the new generation of plastics. The goal in this handbook is to bring together some of the core knowledge in the field to provide a practical and wide-ranging guide for engineers, product designers, and scientists involved in the commercial development of biopolymers and their use in various biomedical and environmental applications.

This book provides information about polymeric biomaterials: plant-derived polymers, methods of manufacture, applications, and disposal. The chapters describe biodegradable and bio-based polymers and plant polymer resources, demand, and sustainability. The book also covers the applicability of chitosan, starch-based biopolymers, and bacterial cellulose in biomedical and environmental applications.

The editors wish to thank the authors, who have generously contributed to this book. They are experts in their fields and provide valuable information and insights into the polymers of the future.

Shakeel Ahmed
Suvardhan Kanchi
Gopalakrishnan Kumar

Chapter 1

Biopolymers: Classification and Applications

Arivalagan Pugazhendhi,[a] Karuppusamy Indira,[b] Jaya Mary Jacob,[c] Malavika Mukesh,[c] and Gopalakrishnan Kumar[d]

[a]*Innovative Green Product Synthesis and Renewable Environment Development Research Group, Faculty of Environment and Labour Safety, Ton Duc Thang University, Ho Chi Minh City, Vietnam*
[b]*Corrosion Science and Technology Division, Indira Gandhi Centre for Atomic Research, Kalpakkam 603102, Tamil Nadu, India*
[c]*Department of Biotechnology and Biochemical Engineering, Sree Buddha College of Engineering, Pattoor, Kerala, India*
[d]*Institute of Chemistry, Bioscience and Environmental Engineering, Faculty of Science and Technology, University of Stavanger, Box 8600 Forus, 4036 Stavanger, Norway*

arivalagan.pugazhendhi@tdtu.edu.vn, gopalakrishnan.kumar@uis.no
gopalakrishnanchml@gmail.com

Biopolymers are recognized as promising candidate materials for diverse field applications due to their outstanding physical, chemical, and mechanical properties. In this chapter, various classifications of biopolymers, from natural to synthetic, have been described according to their chemical characteristics. Polyhydroxyalkanoates (PHA) have been highlighted as an important industrial biopolymer. Apart from that, various applications of biopolymers in different fields such as medicine, drug delivery, and other facets have been discussed. Finally, a

Handbook of Biopolymers: Advances and Multifaceted Applications
Edited by Shakeel Ahmed, Suvardhan Kanchi, and Gopalakrishnan Kumar

ISBN 978-981-4800-17-4 (Hardcover), 978-0-429-02475-7 (eBook)
www.panstanford.com

summary of the future directions have been given for sustainable future with polymer industries.

1.1 Introduction

Biopolymers are recognized as entities whose existence traces back to the age of the first-ever life form on earth. However, the significance of these entities as biodegradable and efficient raw materials has been acknowledged only in the recent past. Ever since, researchers have unanimously focused on harnessing the superior properties of these magic polymers. The term biopolymer stands close to nature in every literal sense—biopolymer meaning biologically derived and biodegradable polymers. It is this "green" factor that has brought about an ardent scientific and research interest toward biopolymers. In the fast growing scenario in science, technology, and medicine, biopolymers have acquired a significant acclaim. They are the renewable raw materials that are essential for the living beings, which possess excellent characteristics and have vital role in many industrial operations. Renewability, sustainability, nontoxic, non-immunogenic, and biodegradability are some of the most important properties of biopolymers that make these materials suitable for multifaceted applications. This chapter deals with the classification and applications of biopolymers.

1.2 Classification of Biopolymers

Figure 1.1 shows the schematic representation of the classification of biopolymers with examples. Accordingly, biopolymers are classified into two major categories: natural and synthetic biopolymers. The advance in technology gives rise to new polymer complexes with unique properties in the desired applications of polymer engineering. The innate ability of many scientists to work on specific parts such as structural, mechanical, and biochemical parts led to the discovery of cellulose, chitosan, starch, PHAs, etc., and their synthesis [1]. The following sections deal with these two types of biopolymers in detail.

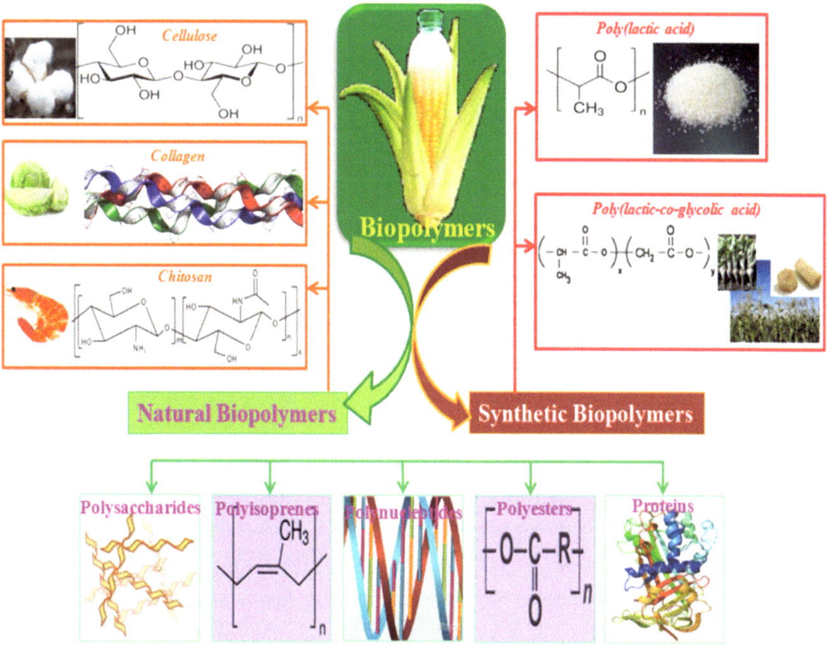

Figure 1.1 Schematic representation of classification of biopolymers with examples.

1.2.1 Natural Polymers

Natural polymers are polymers produced from living organisms and otherwise known as polymeric biomolecules. Plant-originated polymers such as starch, soy protein, and cellulose; animal-originated polymer such as chitosan, keratin, and silk; and microbial fermented polymers such as polyhydroxyalkanoates (PHA) and polyhydroxybutyrate (PHB) are the three renewable resources from which natural polymers can be obtained [2]. The natural biopolymers in the following sections are discussed under the following categories: polysaccharides, polyisoprenes, polynucleotides, polyesters, and proteins [3].

1.2.1.1 Polysaccharides

Polysaccharides fall into the category of carbohydrates and are typically composed of monomer units called monosaccharides linked by glycosidic bonds. Well-known examples of polysaccharides are glucose, cellulose, chitin, etc. These moieties can fall into three groups based on their function, viz., structural polysaccharides (cellulose, chitin, etc.), storage polysaccharides (starch, glycogen, etc.), and gel-forming polysaccharides (agar). Majority of the polysaccharides are located in plants, in which they serve as reinforcement phases. Polysaccharides are also synthesized by algae, tunicates, and some bacteria [4]. Chitin, obtained from the exoskeleton of arthropods, is also seen in the cell wall of fungi and other yeast. It is also produced by a number of living organisms in the lower plant and animal kingdoms and also plays a major role where reinforcement and strength are required [5].

Microbial polysaccharides have recently emerged as important novel biopolymers, which include glucans, pullulan, xanthan gum, and alginase. The unique rheological properties are based on their high purity and regular structure. Microbial polysaccharides are used as thickening and stabilizing agents in food industries. Almost all polysaccharides are pure and are associated covalently or non-covalently with other polysaccharides, polyphenolics, or proteins.

1.2.1.2 Polyisoprenes

Natural rubber with native thermosetting properties constitutes polyisoprene [6], which is classified into two types: *cis* and *trans*. Polyisoprene is also found as Z-polyisoprene and E-polyisoprene. The two forms have different properties. For instance, the mechanical properties of the *cis*-polyisoprene product from rubber tree render high commercial value to the same. The *trans* form with inherent properties such as rigidity, low coefficient of thermal expansion/contraction, acid alkali resistance, and insulation properties is also important from the commercial viewpoint. This form is readily obtained from plants such as *Mimusops balata*. Basically, the isoprene units present in the two forms are responsible for the difference in the functional properties of the two forms of polyisoprene. The *cis* form is more readily available as it is produced by most of the latex producing higher plants, especially rubber tree.

1.2.1.3 Polynucleotides

Polynucleotide biopolymers are formed from long chains of nucleotides. Nucleotides such as ribonucleic acid consist of a nitrogen base, a carbohydrate residue, and a phosphate group linked via phosphodiester bonds. Well-known examples of polynucleotides are acids formed by ribonucleotides (RNA) and deoxy ribonucleic acid (DNA) made of deoxyribonucleotides. A typical dinucleotide molecule carries a free phosphate group at the 5′ end and a free OH-group at the 3′ end. These reactive ends can aid efficient polymerization to form polynucleotides. Usually, the monomerization of around 13 nucleotides could be called a polynucleotide. In a cell, the synthesis of nucleic acids is carried out with the help of enzymes, with the original DNA molecule acting as the template. In practice, polynucleotides are suited to assist engineering in three dimensions by aiding the scaffolding and ordering of materials [7].

1.2.1.4 Polyesters

Structurally, polyesters can be visualized as numerous esters within the fibers whose backbones are formed by the esterification and condensation of poly-functional alcohols and acids. Chemically, these polymers constitute at least 85% of an ester, a dihydric alcohol, and a terephthalic acid by weight. In other words, it is the linking of several esters within the fibers. The best-known example of natural polyesters is the bacterial polyester, known as polyhydroxyalkanoates. PHAs have been well known as osmotically inactive energy storage compounds and have attracted interest because of their biodegradable nature. Generally polyesters are attributed with superior strength and durability. These biopolymers are extremely resistant to chemicals. They also have high abrasion resistance and resist shrinking, stretching, and wrinkling. Polyesters do not absorb water and dry quickly and hence are used as an insulator in the manufacture of hollow fibers [1].

Polyesters are classified into saturated and unsaturated polyesters. The backbone of the polyester family consists of alkyl thermosetting resins based on vinyl unsaturation and is known as unsaturated polyesters. Unsaturated polyesters are widely used in plastic reinforcement and in the economic family of resins. The family of polyesters with a saturated ester backbone that is less reactive than unsaturated counterparts is termed as saturated polyesters. Examples include low-molecular-weight liquids, which are used as plasticizers, as reactants in the production of urethane polymers, and in the production of a linear, high-molecular-weight thermoplastic such as polyethylene terephthalate.

1.2.2 PHAs

The PHA group is mainly divided into two categories: short-chain-length and medium-chain-length polyhydroxyalkanoates. This depends on the number of carbon atoms in the branching polymer chain that vary from 3 to 14 carbon atoms and the homopolymers and heteropolymers making monomeric units [8]. Short-chain-length PHAs (PHASCL) are composed of three to five carbon atoms. Examples of this class are poly(3-hydroxybutyrate) P(3HB), poly(4-hydroxybutyrate) P(4HB) and poly(3-hydroxyvalerate) P(3HV), and copolymer P (3HB-co-3HV). Medium-chain-length PHAs (PHAMCL) are composed of 6–14 or more than 14 carbon atoms. Examples of this class are homopolymers poly(3-hydroxyhexanoate) P(3HHx), poly(3-hydroxyoctanoate)P(3HO), and copolymers such as P(3HHx-co-3HO). The major difference between the two classes is mainly due to the peculiarity of the substrate in PHA synthases that can take 3HAs to a particular range of carbon length. For example, PHA synthase of *Alcaligenes eutrophus* can polymerize 3HAs, which contain three to five carbon atoms, whereas the PHA synthase of *Pseudomonas oleovorans* can only take 3HAs of 6–14 carbon atoms. Hybrid polymers contain both short-chain and medium-chain monomeric units such as poly(3-hydroxybutyrate-co-3-hydroxyhexanoate) [8]. Due of the stereoselectivity of the biosynthetic enzymes, all monomers are in R-configuration. Monomers with different functional groups, such as halogen, hydroxyl, epoxy, cyano, carboxyl, and esterified carboxyl groups have been detected on the chain of mcl-PHAs.

1.2.2.1 Applications of PHAs

Industrial PHAs have a large scope of applications (Table 1.1). In the beginning, they were used as packing materials such as films in bags, cosmetics containers, shampoo bottles, cups, and milk cartons and as latex to cover cardboards and paper. They also include disposable items such as razors, utensils, diapers, feminine hygiene products, pens, combs, bullets, sanitary towels, in molding, golf tees, hot-melt and pressure-sensitive adhesives and also food additives, dairy cream substitutes, non-woven fabrics, binders, thermoformed articles. PHAs are also used in making foils and diaphragms. Because of their piezoelectric nature, PHAs can be used for making pressure sensors for keyboards, stretches and instruments for measuring acceleration, material testing, shock wave sensors, lighters, gas lighters, acoustics (microphone, ultrasonic detectors, sound pressure–measuring instruments, oscillators, headphones, loudspeakers), ultrasonic therapy, and atomization of liquids. The gas barrier character of P(3HB-3HV) is used in food packaging and plastic beverage bottles.

As an addition to their potential as a plastic material, PHAs have great advantage as stereo regular compounds, which serve as chiral precursors for the chemical synthesis of optically active compounds. In the field of medicine, PHAs, especially P(3HB), P(3HB-3HV), P(4HB), P(3HO), and P(3HB-3HHx), are frequently used in tissue engineering. The major contribution of PHAs to medicine is seen in the cardiovascular area (pericardial and atrial septal repair patches, scaffolds for regeneration of arterial tissues, vascular grafts, cardiovascular stents, and heart valves). Also, they are mostly used in bone plates, osteosynthetic materials, surgical sutures, dusting powders, wound dressings, nerve conduits, soft tissue repair, dental and maxillofacial treatment, prodrugs, implants, and tablets. They are useful in the slow release of drugs and hormones. In the areas relating to agriculture, PHAs have been used as mulch films, herbicides, insecticides, and bacterial inoculants, which are useful in enhancing nitrogen fixation in plants.

Table 1.1 Challenges and prospects of PHAs

Challenges	Prospects
High cost of production and extraction	Reduction in cost related to the use of substrate for bacteria growth coming from by products or waste materials; increase PHA production by use of mixed culture or modified bacteria or microalgae; optimization of PHA extraction processes
Quality of PHA	Optimization of the quality and uniformity of PHAs produced in mixed culture.
Mechanical properties	Better understanding of PHA kinetics of crystallization and proper choice of additives (nucleating agents, plasticizers) to achieve stability in mechanical properties and improvement in elongation at break)
Production of blends and composites	Optimization in the use of PHAs in blends with other biodegradable polymers achieving a reduction in cost of the final product while maintaining the outstanding properties of PHAs in terms of barrier properties, modulus, and high biodegradability in different environments.
Blends with natural additives	PHAs in processing are very sensitive to water presence, but proper drying of natural additives and proper choice of compatibilizers are promising for the preparation of blends of PHAs and natural polymers (starch, proteins, etc.) which can achieve the production of plastic items with high biodegradability, for example, also in marine environment

1.2.2.2 Blends of PHAs

Various amalgams of PHAs have been developed to lower the prices and to modify the act of PHAs. PHA/PLA amalgam is the most studied blend as PHB and its copolymer poly(3-hydroxybutyrate-co-hydroxyvalerate) PHBV are really close to PLA in chemical structure and processing [9]. Amalgams of PHA with rubber [10], poly(ethylene glycol) (PEG), poly(ethylene oxide) (PEO), poly(vinyl phenol), polycaprolactone, and poly(propylene carbonate) have also been discovered. Blends of PHA with natural biodegradable polysaccharides, such as cellulose, starch, and their derivatives have also been reported. Morphological and molecular properties of polyhydroxybutyrate-hydroxyvalerate (PHB-HV) with maize starch blends were explored by Reis et al. [11]. The nature of the crystallization process and intermolecular hydrogen bonding interactions of poly(3-hydroxybutyrate) (PHB)/chitin blends were studied [12]. Novel hybrid fibrous membranes along with their potential application in the field of biomedicine were successfully refined through the electrospinning of blend solutions of two biocompatible and biodegradable polymers, poly(3hydroxybutyrate-co-hydroxyvalerate) (PHBV)/chitosan. In another study, because of the effects of hydrogen bond, the characterization of intermolecular interactions on the crystal spherulite of PHB/CAB blends was done. Environmental deterioration of lignin/PHB blends along with the role of lignin in either decreasing or increasing the PHB deterioration has been explored in this above study.

1.2.2.3 Nanocomposites of polyhydroxyalkanoates

PHBV nanocomposites were recovered by the addition of two different type of nanoparticles, organically altered platelet such as clay montmorillonite Cloisite 30B (C-30B) and unaltered tubular-shaped halloysite, and investigated their effect on the morphology and thermal and mechanical characters of PHBV nanocomposites. A partially desquamated structure for PHBV/C30B nanocomposites and a good scattering of HNT in the PHBV matrix were seen in a study by Carli et al. [13]. Mechanical properties of kenaf and lyocell fiber reinforced poly(3-hydroxybutyrate) composites and were correlated [14]. Poly(3-hydroxybutyrateco-3-hydroxyvalerate)/organo-modified Mg-Al layered double hydroxide (PHBV/m-LDH) and poly(3-hydroxybutyrate) (PHB)/mLDH nanocomposites were synthesized and their structure, morphology, and physical properties were defined [15]. Mechanical properties, bio-compatibility, and biodegradability of composites of treated (cross-linked) cellulose acetate (t-CA) and acrylic acid-grafted poly(hydroxyalkanoate) (PHA-g-AA/t-CA) were explored in a study by Wu [16].

Structure, crystallization behaviors, and thermal properties of poly(3-hydroxybutyrate-co3-hydroxyvalerate)/organophilic montmorillonite (PHBV/OMMT) nanocomposites were also considered. The relationship between the structure and property of silica/poly(3-hydroxybutyrate-co-3-hydroxyhexanoate) (PHBHx) nanocomposites were researched in an article. In this study, biodegradable nanocomposites were prepared with the use of nanofibrillated cellulose (NFC) as the reinforcement and poly(3-hydroxybutyrate-co-3-hydroxyvalerate, PHBV) as the polymer matrix. The mechanical properties, biocompatibility, and biodegradability of composite materials made from chestnut shell fiber (CSF) and poly(hydroxyalkanoate) (PHA), as well as CSF and glycidyl methacrylate grafted PHA (PHA-g-GMA), were checked [16]. Poly(3-hydroxybutyrate-co3-hydroxyvalerate) based nanocomposites that consist of a variable quantity of organophilic montmorillonite were made using the solution intercalation method. Thermal stability, tensile properties, and nanomechanical material properties were also researched by the nanoindentation method. Poly(3-hydroxybutyrateco-3-hydroxyvalerate) (PHBV)/cellulose nanowhisker (CNW) composite with unidirectional CNW orientation were arranged. CNW in poly(3-hydroxybutyrate-co-3-hydroxyvalerate) (PHBV) matrix was aligned using an external electric field, and the morphology and mechanical properties of the resultant anisotropic composites were studied [17]. Thermophysical properties and biodegradation behavior of green composites made from PHB and potato peel waste fermentation residue (PPW-FR) fibers were investigated [18]. Nanocomposites of poly(3-hydroxybutyrate-co-3-hydroxyhexanoate) (PHBHHx)/layered silicates and PHBHHx/expanded graphite were produced. For the nanofillers used in the two nanocomposites, a desquamated morphology with good distribution at lower nanofiller content was observed. Synthesis of a shape memory polymer nanocomposite derived from poly(3-hydroxyoctanoateco-3-hydroxyundecenoate) composited with nanofillers of silsesquioxane (POSS) was observed. A nanocomposite membrane with respect to a PHB-functionalized multi-walled carbon nanotubes/chitosan matrix was used efficiently to pervaporate a mixture of water and 1,4-dioxane [19].

1.2.3 Proteins

Polymers that are formed by a succession of amino acid subunits joined by peptide bonds are known as proteins. Proteins in their natural state are present as either globular or fibrous forms. The globular proteins fold into complicated sphere-shaped structures held by a mutual arrangement of hydrogen, ionic, hydrophobic, and covalent (disulphide) bonds. The fibrous proteins are completely expanded and strongly coupled in lateral constructions by hydrogen bond to form fibers. The chemical and physical properties of these proteins depend on the equivalent volume of amino acid residues and their placement along the polypeptide chain.

Keratin is a natural protein that contains disulfide bonds and has an array of properties that vary from a structurally vigorous and shock-resistant material (horn) to a simple waterproof layer (turtle shell). Keratin is mechanically efficient in tension (wool) and compression (hooves). Keratin is normally found in hair, wool, claws, nails, skin, fur, hooves, beaks, feathers, horns, scales, actin, and myosin protein, which is found in muscle tissues. Silk is an outstanding biological polymer but has a complex structure. Silk are produced by some insects such as spider and silkworm (Bombyxmori), but mostly the silks of larvae have been used for casting textiles. Most of the silks have extraordinary mechanical characters and exhibit an incomparable combo of high tensile strength and extensibility. The arrangement of firmness and resilience provides silks high toughness, which is equivalent to commercial aromatic nylon filaments, which are benchmarks of current polymer fiber technology. Spider silk has been documented in long run as a miraculous fiber for its peculiar combo of high strength and break elongation [2].

1.2.4 Synthetic Biopolymers

Synthetic biopolymers are classified based on their preparation method, especially the addition and polymerization reactions. Scientists and engineers derived synthetic polymers from petroleum oil. Commercially important examples of synthetic polymers include nylon, polyethylene, polyester, Teflon, and epoxy. A few important basic examples of synthetic polymers discussed in this chapter are poly(lactic-co-glycolic acid) (PLGA) and poly(lactic acid) (PLA).

1.2.4.1 Poly(lactic-co-glycolic acid)

Poly(lactic-co-glycolic acid) PLGA is widely being used as a biodegradable biopolymer, which is made from inexhaustible resources. As a result of environmental and economic concerns combined with waste disposal and increase in costs of petroleum production, which results from the depletion of the most easily available reserves, biodegradable polymers have become more attractive economically [20]. PLGA is obtained from the reaction of lactic acid with glycolic acid as shown in the following reaction:

1.2.4.2 Poly(lactic acid)

Poly(lactic acid), a biodegradable polymer, is obtained by dehydration and ring-opening polymerization methods resulting in low- and high-molecular-weight PLAs, respectively as shown in the following reactions [21].

Lactic acid, the main compound in PLA, is chiral with two asymmetric centers and exists in four contrasting forms, which make the PLA to have a wide range of properties. This is being made into a polymer with molecular weight ranging from a few thousands to over a million and is easily converted into film, fiber, spun bond, and melt-blown commodities on current processing machineries. Considering its application in the medical field, PLA has practical operations as dissolvable sutures and as matrices for drug delivery and bone fracture internal fixation devices in surgery. It is also used as an agricultural plant growth promoter, in textiles, and non-woven functions such as fiberfill, crop covers, geotextiles, wipes, diapers, binder fibers, etc. The content of hydrolysis of PLA correlated with polyvinyl acetate or ethylene-acrylic acid copolymers, which makes PLA a substitute binder for cellulosic non-woven fiber. Initially PLA was incorporated about 50 years ago and has cautiously arrived as a substitute to PET, HIPS, PVC, and cellulosics in a few high-clarity packaging roles. Furthermore, it has been used as an adhesive layer in form-fill-seal co-extrusions, in candy wrap, optically strengthened films, and shrink labels and has forging roles in thermoformed cups and containers and will soon appear as a single-serve drinking bottles [20].

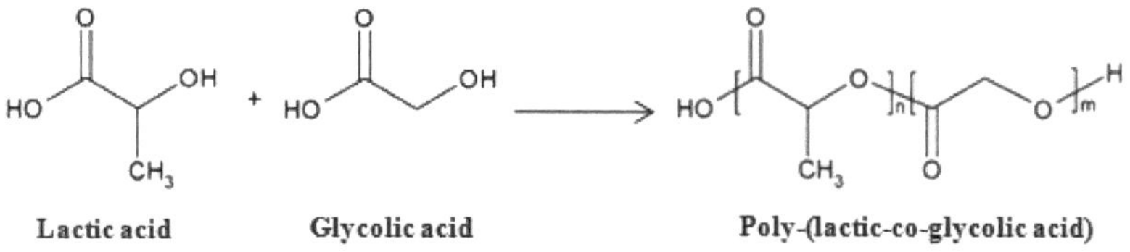

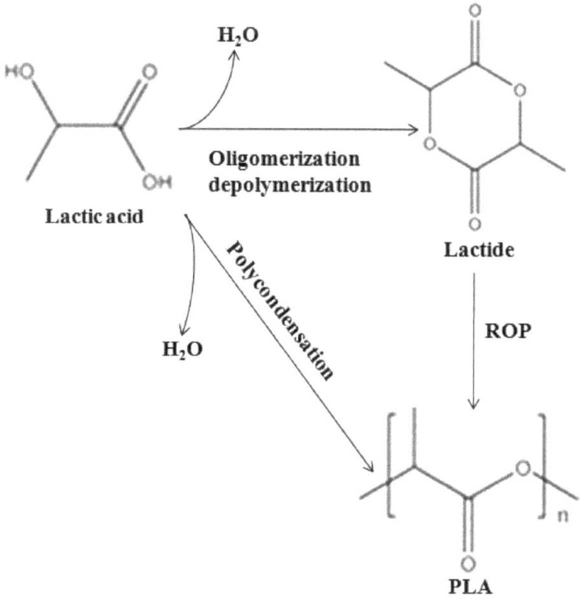

1.3 Applications

Biopolymers literature on the application of biopolymers in packaging, medicine, agriculture, and other areas has begun and various explored applications of biopolymers have been shown (Fig. 1.2).

Biopolymers possess certain characteristics such as biocompatibility, biodegradation to non-toxic end products, high bio-activity, low anti-genicity, capability to support cell growth and proliferation with correct mechanical operations, ability to process complex shapes with apt porosity, and preserving mechanical firmness, which make them a suitable material in the field of medicine, such as drug delivery systems, surgical implant devices, and bruise closing and healing products. Likewise, biopolymers are mostly used in food containers, soil retention sheeting, agriculture film, and waste bags and are also used in packaging material due to their film-forming and barrier properties. Additionally, they are popular in areas such as automotive development, hazardous waste removal, paper industries, and development of new building materials.

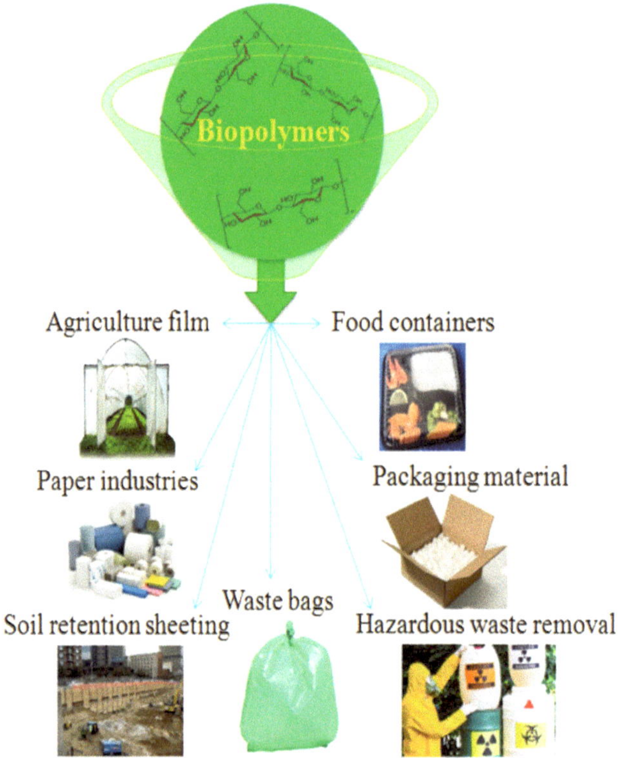

Figure 1.2 Schematic representations of applications of biopolymers.

1.3.1 Medicine

In the global medicinal market, biopolymers have found invaluable applications. The demands for natural biopolymers are quite high because of their unique properties and growing areas of application in medicine. The following sections highlight some of the important application of biopolymers in recent times.

1.3.1.1 Drug delivery systems

A substantial challenge for the scientist in the biomedical field is the advancement of relevant carriers for the delivery of drugs. In order to overcome this challenge, deliveries of drugs in a controlled, degradable, and safe manner have attracted increasing importance in recent years, resulting in a remarkable progress in their development [22].

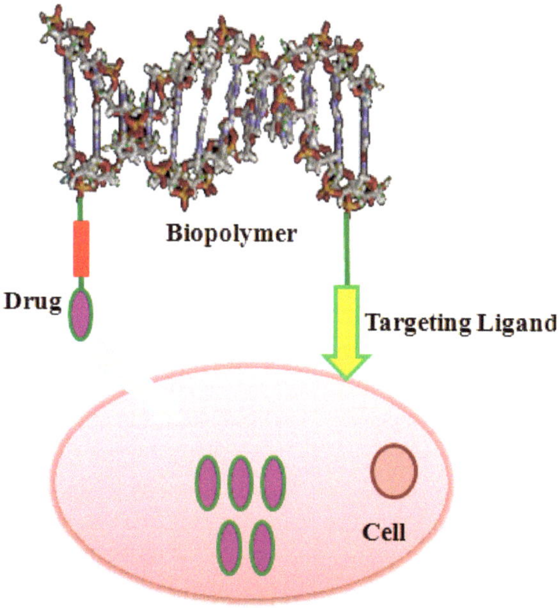

Figure 1.3 Schematic representation of biopolymer–drug conjugation with target cell-specific ligand.

Biocompatible and biodegradable natural polymer classes such as proteins, polysaccharides, and natural polyesters are the most studied biopolymer in the biomedical field, which are readily available from natural sources and accommodate definite signals for cell adhesion, allowing cell aggression. For designing innovative nanosystems to delivery bioactive compounds into a specific site, in a controlled manner the aforesaid features highlight their applications. The efficiency, specificity, acceptability, and bioavailability of bio-actives can be improved in this manner. In recent years, research has been carried out for the advancement in the new drug delivery systems for targeted delivery of drugs using biopolymers. Figure 1.3 shows the schematic representation of biopolymer–drug conjugation with target cell-specific ligand, which offers a solution for selective and targeted delivery.

Biopolymers can be tailored as drug delivery carriers due to their targeting ability, non-immunogenicity, and biodegradability. Nanoparticle structures of biopolymers have been desired as carriers for systemic and targeted drug delivery. Nanoparticles provide inherent drug stabilization, which significantly improves the pharmacokinetics of the drug. In this context, chitosan biopolymer has enormous structural prospects for chemical and mechanical alterations to generate novel characters, behavior, and operations, especially in the biomedical field. It is an acceptable candidate for traditional and unique delivery of drugs because of its outstanding characters such as non-toxic, biocompatibility, and biodegradability. It produces colloidal particles, which entangle bioactive molecules over a number of mechanisms, combining chemical cross-linking, ionic cross-linking, and ionic complexation. It can be readily processed into gels, sponges, membranes, beads, and scaffolds models [23].

As chitosan microspheres can settle longer in the stomach, they can also be used for stomach-specific delivery of drug. They are also used in colon-specific delivery of drug as they are depraved by the microbes found in the colon. In addition, they are used as a promising candidate for oral cavity drug delivery and are a convenient material for powerful non-viral gene and DNA vaccine delivery. Chitosan has been worn in many contrasting establishments in drug and gene delivery in the gastrointestinal tract, which is a result of its excellent physical, chemical, and biological properties [24]. It can be incorporated with drugs such as fluriprofen and celecoxib, which can be used as a carrier for intra-articular agents to prolong the confinement of the drug in the knee joint [25, 26].

Drug delivery for fundamental effects over skin in a prearranged and restrained rate can be achieved by transdermal delivery of drugs in the form of hydrogel sheaths, which avoids the first-pass metabolism effect. Drug delivery can be effortlessly broken up on demand by simply removing the devices. When ophthalmic drugs are administered in hydrogels, the association time of drugs with cornea has been increased resulting in the increases of ocular bioavailability. Besides, in the in situ thermosensitive hydrogel composed of chitosan and b-glycerophosphate, the hydrogel improved the transcorneal permeation to sevenfold over an aqueous solution and enhanced ocular bioavailability. Ocular delivery of timolol maleate drug release from hydrogel was also doubled if the in situ thermosensitive hydrogel was composed of chitosan and poly(N-isolpropylacrylamide) [27].

Dev et al. (2010) [28] described that the combined effect of PLA/chitosan nanoparticles is non-toxic in mouse fibroblast cells and they approve that the PLA/chitosan nanoparticle system can be useful in biomedical applications, including drug encapsulation studies. The rate of deterioration of these nanoparticles has been expanded expeditiously when pH expands in the range of 8–13. Accordingly, such nanoparticles can entangle and armor drugs in the stomach environment at acidic pH and implement extended release in the intestine at neutral pH [28]. The solubility of least soluble drugs can be expanded by hydrophobically modified chitosan derivatives. The four major classes of hydrophobically altered chitosan are steroid derivatives, fatty acids derivatives, aryl and alkyl derivatives, and carboxymethyl derivatives of chitosan, which are dynamically used as a potential drug delivery carrier for least soluble drugs.

1.3.1.2 Surgical implants

The biopolymers are the best choice for surgical implant applications as implantable medical devices constituted by biodegradable polymers do not require surgical removal. For tissue engineering applications, where polymeric scaffolds are designed to support and promote cell growth to regenerate a particular damaged tissue, these are the relevant advantages.

1.3.2 Agriculture

The global population has been increasing at an alarming rate, and by the end of year 2050, it is expected to reach 7 billion. The priority of the agricultural evolution all over the world during the 20th century was constantly expanding fertility per unit area of land used for growing crops to feed the increasing population. Natural water resources were exploited and were contaminated with hazardous pesticides and chemical fertilizers; hence, the fertility of the cultivable land was impaired, which was critical for the survival and welfare of all life on earth. Consequently, the priority on agricultural advancement in the 21st century has shifted to the sustainable use of land, water, and plant resources. Thus, the most important goal of the present-day agriculture is to boost land and water productivity without harming the environment and its natural resources. In the current situation, the use of polymers in agriculture is achieving popularity in science, which has already provided solutions to the present-day agricultural problems. Soil permeability, density, structure, texture, evaporation, and infiltration rates of water through the soils have been likely changed by superabsorbent polymer hydrogels. The functionalized use of polymers has increased the efficiency of pesticides and herbicides by using low doses that eventually safeguard the environment by dropping the pollution level and cleaning up the existing pollutants [29].

Low erratic rainfall, high evaporation Level, and long periodic droughts were resulting in arid and semi-arid soil. The water quality and agricultural productivity were threatened by soil deterioration by the loss of prized top soil. Perhaps, biopolymers have been identified as achievable soil conditioners, as they maintain the structure of soil surface and pore durability. Water-soluble polymers were high effective in preventing erosion and aggression building in the past decade [30].

1.4 Conclusion

This chapter presents the classification and emerging applications of biopolymers in various field, particularly, biomedicine, energy, and environmental research. In drug delivery systems and surgical implant applications, biopolymers have gained increasing importance in recent times. PHA is an invaluable polymer for a wide range of applications. For example, it has improved boundary properties (OTR and WVTR) and mechanical durability compared with other common bioplastics such as PLA. In spite of its innate fragility, the great development has been presented here, which has been made freshly by the production of PHB with tailored supplements and amalgams leading to highly upgraded mechanical profiles, as well as convenient process ability via extrusion or injection and molding among others. These approaches will reform its capacity to pass through markets such as food packaging. Although limitations still remain, such as the high cost of commercial PHA, low market availability, and the use of refined/ food challenging feedstock. As various research studies are being carried out regarding the advancement in the yield of PHA by genetic modification of bacteria or by the use of waste for their growth. Lastly, more advances could allow even more flexible grades of PHAs or clear ones over the control of its crystallization.

Acknowledgements

This work was supported by the Korea Research Fellowship Program through the National Research Foundation of Korea (NRF) funded by the Ministry of Science and ICT (Grant No: 2016H1D3A1908953).

References

1. Papageorgiou, G. Z., Papageorgiou, D. G., Terzopoulou, Z., and Bikiaris, D. N. (2016). Production of bio-based 2,5-furan dicarboxylate polyesters: Recent progress and critical aspects in their synthesis and thermal properties. *Eur. Polym. J.*, **83**, pp. 202–229.
2. Gupta, P., and Nayak, K. K. (2015). Characteristics of protein-based biopolymer and its application. *Polym. Eng. Sci.*, **55**, pp. 485–498.
3. Tirrell, J. G., and Tirrell, D. A. (1996). Synthesis of biopolymers: Proteins, polyesters, polysaccharides and polynucleotides. *Curr. Opin. Solid State Mater. Sci.*, **1**, pp. 407–411.
4. Sahraee, S., Milani, J. M., Ghanbarzadeh, B., and Hamishehkar, H. (2017). Physicochemical and antifungal properties of bio-

nanocomposite film based on gelatin-chitin nanoparticles. *Int. J. Bio. Macromol.*, **97**, pp. 373–381.

5. Ramaprasad, A. T., Latha, D., and Rao, V. (2017). Synthesis and characterization of polypyrrole grafted chitin. *J. Phys. Chem. Solids.*, **104**, pp. 169–174.
6. Takano, A., Horaiya, T., Odamaki, F., Akazawa, Y., Ohta, Y., Kawaguchi, D., Takahashi, Y., and Matsushita, Y. (2012). Preparation and characterization of polyisoprenes and polybutadienes having 1,2- and 3,4-linkages preferentially. *Polymer*, **53**, pp. 3354–3359.
7. Correia-Pinto, J. F., Peleteiro, M., Csaba, N., González-Fernández, Á., and Alonso, M. J. (2015). Multi-enveloping of particulated antigens with biopolymers and immunostimulant polynucleotides. *J. Drug Deliv. Sci. Technol.*, **30**, pp. 424–434.
8. Taguchi, S., and Doi, Y. (2004). Evolution of polyhydroxyalkanoate (PHA) production system by "enzyme evolution": Successful case studies of directed evolution. *Macromol. Biosci.*, **4**, pp. 146–156.
9. Li, L., Huang, W., Wang, B., Wei, W., Gu, Q., and Chen, P. (2015). Properties and structure of polylactide/poly (3-hydroxybutyrate-co-3-hydroxyvalerate) (PLA/PHBV) blend fibers. *Polymer*, **68**, pp. 183–194.
10. Bhatt, R., Shah, D., Patel, K. C., and Trivedi, U. (2008). PHA–rubber blends: Synthesis, characterization and biodegradation. *Bioresour. Technol.*, **99**, pp. 4615–4620.
11. Reis, K. C., Pereira, J., Smith, A. C., Carvalho, C. W. P., Wellner, N., and Yakimets, I. (2008). Characterization of polyhydroxybutyrate-hydroxyvalerate (PHB-HV)/maize starch blend films. *J. Food Eng.*, **89**, pp. 361–369.
12. Khasanah, Reddy, K. R., Sato, H., Takahashi, I., and Ozaki, Y. (2015). Intermolecular hydrogen bondings in the poly(3-hydroxybutyrate) and chitin blends: Their effects on the crystallization behavior and crystal structure of poly(3-hydroxybutyrate). *Polymer*, **75**, pp. 141–150.
13. Carli, L. N., Crespo, J. S., and Mauler, R. S. (2011). PHBV nanocomposites based on organomodified montmorillonite and halloysite: The effect of clay type on the morphology and thermal and mechanical properties. *Composite Part A: Appl. Sci. Manufactur.*, **42**, pp. 1601–1608.
14. Graupner, N., and Müssig, J. (2011). A comparison of the mechanical characteristics of kenaf and lyocellfibre reinforced poly(lactic acid)

(PLA) and poly(3-hydroxybutyrate) (PHB) composites. *Composites Part A: Appl. Sci. Manufact.*, **42**, pp. 2010–2019.

15. Ciou, C.-Y., Li, S.-Y., and Wu, T.-M. (2014). Morphology and degradation behavior of poly(3-hydroxybutyrate-co-3-hydroxyvalerate)/layered double hydroxides composites. *Eur. Polym. J.*, **59**, pp. 136–143.
16. Wu, C.-S. (2014). Mechanical properties, biocompatibility, and biodegradation of cross-linked cellulose acetate-reinforced polyester composites. *Carbohydr. Polym.*, **105**, pp. 41–48.
17. Ten, E., Jiang, L., and Wolcott, M. P. (2013). Preparation and properties of aligned poly(3-hydroxybutyrate-co-3-hydroxyvalerate)/cellulose nanowhiskers composites. *Carbohydr. Polym.*, **92**, pp. 206–213.
18. Wei, L., Liang, S., and McDonald, A. G. (2015). Thermophysical properties and biodegradation behavior of green composites made from polyhydroxybutyrate and potato peel waste fermentation residue. *Ind. Crops Prod.*, **69**, pp. 91–103.
19. Ong, Y. T., Ahmad, A. L., Zein, S. H. S., Sudesh, K., and Tan, S. H. (2011). Poly(3-hydroxybutyrate)-functionalised multi-walled carbon nanotubes/chitosan green nanocomposite membranes and their application in pervaporation. *Sep. Purif. Technol.*, **76**, pp. 419–427.
20. Mehta, R., Kumar, V., Bhunia, H., and Upadhyay, S. N. (2005). Synthesis of poly(lactic acid): A review. *J. Macromol. Sci. Part C,* **45**, pp. 325–349.
21. Berezina, N., and Martelli, S. M. (2014). Bio-based polymers and materials. *Renew. Resources Biorefineries*, RSC Publishing, UK, pp. 1–28.
22. Indira, K., Mudali, U. K., Nishimura, T., and Rajendran, N. (2015). A review on TiO_2 nanotubes: Influence of anodization parameters, formation mechanism, properties, corrosion behavior, and biomedical applications. *J. Bio. Tribo. Corros.,* **1**, p. 28.
23. Rajan, M., Raj, V., Rajan, M., and Raj, V. (2013). Potential drug delivery applications of chitosan based nanomaterials. *Int. Rev. Chem. Eng.*, **5**, pp. 145–155.
24. Prabaharan, M., and Mano, J. F. (2005). Chitosan-based particles as controlled drug delivery systems. *Drug Deliv.*, **12**, pp. 41–57.
25. Thakkar, H., Sharma, R. K., Mishra, A. K., Chuttani, K., Murthy, R. S. R. (2004). Celecoxib incorporated chitosan microspheres: In vitro and in vivo evaluation. *J. Drug. Target.*, **12**, pp. 549–557.

26. Janssen, M., Mihov, G., Welting, T., Thies, J., and Emans, P. (2014). Drugs and polymers for delivery systems in OA joints: Clinical needs and opportunities. *Polymer*, **6**, pp. 799–819.

27. Giri, T. K., Thakur, A., Alexander, A., Ajazuddin, Badwaik, H., and Tripathi, D. K. (2012). Modified chitosan hydrogels as drug delivery and tissue engineering systems: Present status and applications. *Acta Pharm. Sin. B*, **2**, pp. 439–449.

28. Dev, A., Binulal, N. S., Anitha, A., Nair, S. V., Furuike, T., Tamura, H., and Jayakumar, R. (2010). Preparation of poly(lactic acid)/chitosan nanoparticles for anti-HIV drug delivery applications. *Carbohydr. Polym.*, **80**, pp. 833–838.

29. Ekebafe, L. O., Ogbeifun, D. E., and Okieimen, F. E. (2011). Polymer applications in agriculture. *Biokemistri*, **23**, pp. 81–89.

30. Maghchiche, A., Haouam, A., and Immirzi, B. (2010). Use of polymers and biopolymers for water retaining and soil stabilization in arid and semiarid regions. *J. Taibah Univer. Sci.*, **4**, pp. 9–16.

Chapter 2

Bio-Based Biopolymers and Their Potential Applications for Bio- and Non-Bio Sectors

Muhammad Bilal[a] and Hafiz M. N. Iqbal[b]

[a]*State Key Laboratory of Microbial Metabolism, School of Life Sciences and Biotechnology, Shanghai Jiao Tong University, Shanghai 200240, China*
[b]*Tecnologico de Monterrey, School of Engineering and Sciences, Campus Monterrey, Ave. Eugenio Garza Sada 2501, Monterrey, N. L., CP 64849, Mexico*

bilaluaf @hotmail.com, hafiz.iqbal@itesm.mx

This chapter focuses on the contemporary research trends, scientific developments, and future perspective of biopolymers and their potential applications for bio- and non-bio sectors. In the recent years, bio-based and biodegradable polymers have attracted increased research attention owing to environmental apprehensions, confined, boundaries and vagueness in supplies of the fossil. Bio-based polymers not only substitute synthetic materials in a variety of different applications but also provide some tailor-made novel characteristics for targeted applications

Handbook of Biopolymers: Advances and Multifaceted Applications
Edited by Shakeel Ahmed, Suvardhan Kanchi, and Gopalakrishnan Kumar

ISBN 978-981-4800-17-4 (Hardcover), 978-0-429-02475-7 (eBook)
www.panstanford.com

by carrying out derivatization reactions on the polysaccharide backbone. As functional materials, bio-based polymers, alginate, and chitosan offer a unique set of distinct features, including biocompatibility, biodegradability, non-toxicity, biological inertness, antimicrobial activities, metal-based ions chelation, gelation characteristics, wettability features, and notable affinity to proteins. Chitosan, alginate, and chitosan-alginate-based biopolymers are being anticipated and widely utilized in various sectors of the modern world. The first half of this chapter describes unique physiochemical characteristics of alginate and chitosan. The second part mainly focuses on the potentialities of these materials for environmental engineering and biomedical engineering. In summary, recent progress in bio-based polymers, in particular, alginate and chitosan, has been presented in this chapter, with respect to the structure, general properties to their potential biomedical and biotechnological applications.

2.1 Introduction

Bio-based polymers derived from natural origins are one of the omnipresent classes of materials in biomedical and biotechnology industry. Owing to the current demands and necessity, a profound attention in biopolymers has accelerated, across the globe, with an aim to find petro- and chemo-free alternatives. In this context, biopolymers offer noteworthy possibilities by reducing the global petro-dependence along with positive ecological influences such as reduced carbon dioxide (CO_2) discharges [1, 2]. In the scientific literature, various terminologies such as biopolymers or bio-based polymers and biodegradable polymers have been widely utilized [3–10]. However, there is a key discrepancy among these polymers. Among them, biodegradable polymers are defined as "materials whose physical/chemical properties undergo decomposition and completely degraded to carbon dioxide, methane, and water by the enzymatic action of microorganisms." On the other hand, bio-based polymers can be biodegradable or non-degradable. Keeping this in view, most of biopolymers or bio-based polymers are biodegradable, e.g., polyhydroxyalkanoates (PHAs) [3–8]. However, not all biodegradable polymers are necessarily considered bio-based polymers, e.g., polycaprolactone [11]. The UN World Commission on "Environment and Development in Our Future" describes sustainability as "the development which meets the necessities of the present-day without negotiating the ability of future generations to encounter their needs." According to this concept, the products must be designed, manufactured and manipulated from "conception to reincarnation," the so-called "cradle-to-grave" approach [1].

The first-generation bio-based polymers are derived from lignocellulose-based materials, including various agricultural or agro-industrial wastes materials such as corn, potatoes, and other carbohydrate foodstuffs. In recent years, the focus, however, has diverted due to direct competition with the world's food supply chains and a fascination for moving away from food-based resources. Nowadays there has been a growing interest in producing biopolymers-based bioproducts through microbial fermentation technologies by taking advantage of renewable resource–based raw materials, lignocellulosic waste biomass, fatty acids, and organic waste [12–15]. These bio-based polymers have been reported to display enormous technological developments and considerable applications in tissue engineering [16]. Figure 2.1 illustrates an overview of the recent uses of biomaterials for biomedical and biotechnological applications.

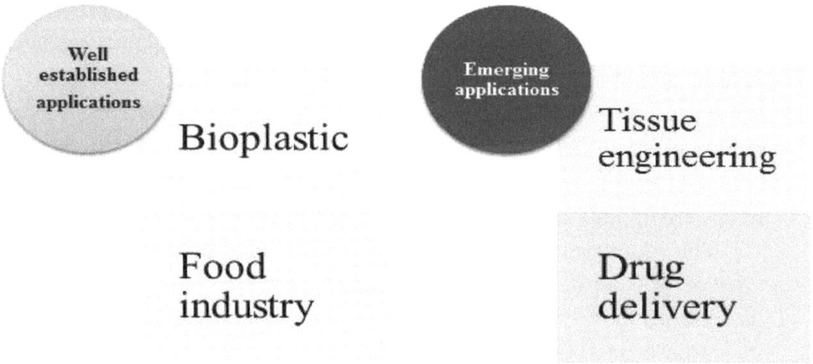

Figure 2.1 Emerging and well-established applications of novel biomaterials or biomaterials-based novel constructs.

2.2 Alginate: General Properties

Alginate is a naturally occurring anionic biopolymer that has been extensively exploited for many biomedical applications, due to its distinctive characteristics such as low toxicity, low cost, biocompatibility, and gelation under an extremely mild environment by the addition of divalent cations such as Ca^{2+} [17]. Commercially available alginates are prevalently extracted from brown algae, including *Ascophyllum nodosum*, *Laminaria digitata*, *Laminaria hyperborea*, *Laminaria japonica*, and *Macrocystis pyrifera* by treatment with aqueous alkali solutions [18]. Following that, alginate is precipitated out by adding either sodium or calcium chloride to the filtered extract. Finally, a range of purification strategies is adopted to obtain a purified and water-soluble sodium alginate power [19]. On the other hand, the bacterial biosynthetic route may provide alginate with relatively more distinct physiochemical structural properties than seaweed-derived alginate. Apart from seaweed-derived alginate, it can be synthesized from microbial cultures, e.g., *Azotobacter* and *Pseudomonas* [20]. The ease of microbial modification via genetic engineering can enable the regulation of alginate biosynthesis in bacteria with desired functionalities. This will additionally support to target the wider application range in biotechnological and biomedical sectors [17]. Regarding composition, alginate is a water-soluble linear polysaccharide copolymer consisting of alternating blocks of (1,4)-linked α-L-guluronate (G) and β-D-mannuronate (M) residues. It is worth noting that the composition, structural characteristics and molecular weight, all play a critical role in governing the unique properties of the alginates. Alginate possesses different physical properties [21]. Figure 2.2 illustrates chemical structure of repeated units of alginate: β-D-mannuronic acid (M) and α-L-guluronic acid (G), and the schematic crosslink between alginate and the counterions [22].

The above-discussed physiochemical characteristics play a critical role while developing alginate-based constructs, and the phenotype and performance of cells/biocatalysts encapsulated in alginate gels. Commercial grade sodium alginate with a wider molecular weight range from 32,000, and 400,000 g/mol is available in sufficient quantity for research purposes. Such availability with a wider molecular weight range offers numerous possibilities to upgrade/modify the physicochemical features of alginate-based engineered constructs including gels. Nevertheless, an alginate solution developed from high molecular weight becomes greatly viscous and gelatinous, which is often undesirable in processing for protein or cell encapsulation [23, 24]. However, the pre-gelling and post-gelling limitations can independently be controlled by manipulating the molecular weight of the biopolymer and its precise distribution in the working environment [25].

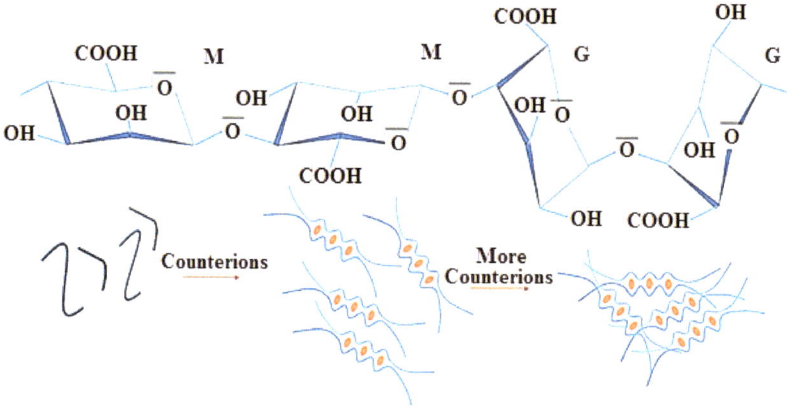

Figure 2.2 Chemical structure of repeated units of alginate: β-D-mannuronic acid (M) and α-L-guluronic acid (G), and the schematic crosslink between alginate and the counterions. Adapted from Ref. 22, an open access article distributed under the terms of the Creative Commons Attribution License (http://creativecommons.org/licenses/by/4.0).

From a regulatory point of view, alginates have been categorized in a group of compounds that are regarded as safe (GRAS) material by the U.S. Food and Drug Administration (FDA). Alginates purified by a multi-step extraction protocol to a very high purity did not tempt any noticeable foreign body reaction when inoculated into animals [26]. Similarly, little or no significant inflammatory responses were provoked when commercial scale highly purified alginate-based gels administered intravenously into mice, thus corroborating its non-toxic and biodegradable nature [27]. Any immunogenic response at intravenous injections might have been experimented much likely due to impurities or toxic contaminants remaining in the commercial alginates [28].

2.3 Chitosan: General Properties

Chitosan is a linear mucopolysaccharide containing β (1–4)-linked 2-amino-2-deoxy-D-glucose (D-glucosamine) and 2-acetamido-2-deoxy-D-glucose (N-acetyl-D-glucosamine) units. The chitin and chitosan structures are diagrammed in Fig. 2.3, where, "*n*" is the degree of polymerization [22]. Given structural analysis, chitosan is very identical to that of cellulose given its solubility and low chemical reactivity. It may be regarded as cellulose in which there are hydroxyl groups at carbon-2 positions of the glucose rings is replaced by an amino group. Chitosan, the catch-all idiom, refers to describe a series of polymers with different degrees of deacetylation. The deacetylation degrees and the molecular weight of commercially available chitosan typically range 70–95% and 10–1000 kDa, respectively. Inarguably, the physical properties, biodegradability, and biological perspective of chitosan are highly reliant on the relative magnitudes of D-glucosamine and N-acetyl-D-glucosamine residues [29].

Unlike alginate, chitosan is a second most abundant biopolymer after cellulose with cationic nature. Chitosan is a polysaccharide that can be produced, in bulk, from chitin found in the exoskeletons of crustaceans at low costs. Figure 2.4 illustrates a schematic presentation of chitosan preparation from raw materials [30]. The α-chitosan derived from crab- and shrimp shell wastes is among the most commonly available form of chitosan [31] and accounts for approximately 70% of the organic compounds in such shells. In recent years, apart from natural sources, considerable attention has been given to fermentation approaches for the biosynthesis of chitosan from fungal mycelia [32]. Its natural abundance, biocompatibility, biodegradability, non-toxicity, and antiviral and antibacterial properties allow this biomaterial and its different combinations with other materials to be an excellent candidate for biomedical applications such as drug delivery systems. Additionally, it has the unique potential to participate as an outstanding polymeric matrix for pharmaceutical carriers or gene delivery systems in diverse morphologies [33]. Moreover, the cationic nature of chitosan is essential for its properties mentioned above and also for the delivery of drugs [34].

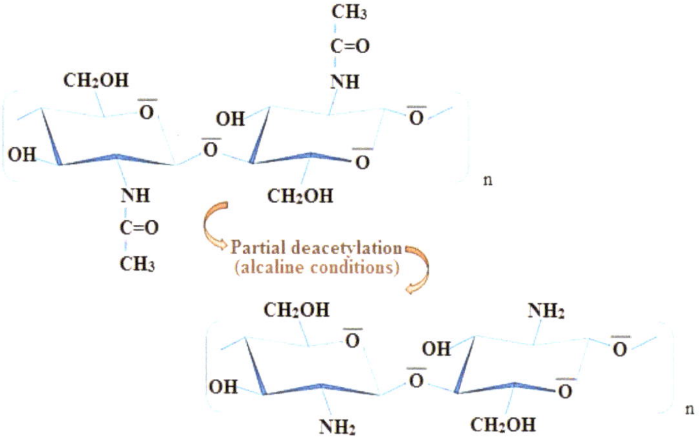

Figure 2.3 Representation of chitin and chitosan structures. Here, "*n*" is the degree of polymerization. Adapted from Ref. 22, an open access article distributed under the terms of the Creative Commons Attribution License (http://creativecommons.org/licenses/by/4.0).

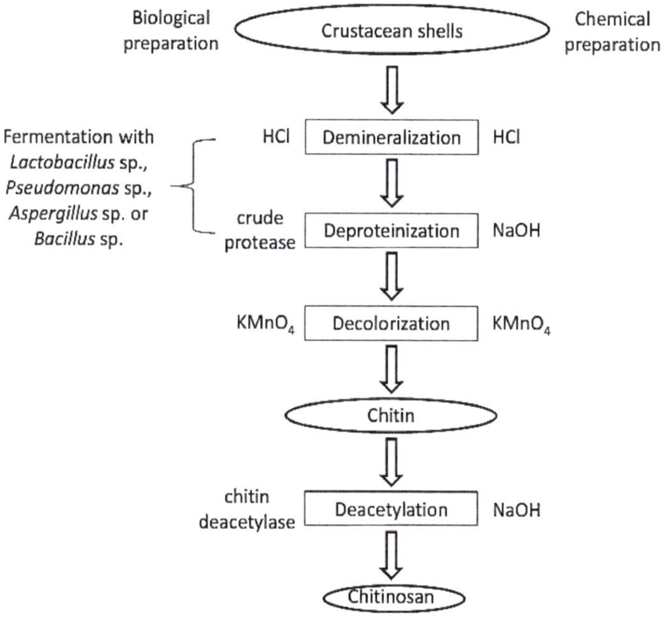

Figure 2.4 A schematic presentation of chitosan preparation from raw materials. Adapted from Ref. 30, an open access article distributed under the terms of the Creative Commons Attribution License (http://creativecommons.org/licenses/by/4.0).

2.4 Biopolymers and Immobilization Engineering

The characteristics of immobilized enzymes are regimented by the properties of both the enzyme and the support material [35]. The appropriate interface between the two confers an immobilized enzymeparticular physicochemical and dynamic properties. This may be fundamental for its applied standpoint, and thus, a support comprehendingly chosen can markedly boost up the functional performance of the engineered system. Although it is demonstrated that no universal support is available for all enzymes and their applications, some anticipated features, including high affinity to proteins, the presence of functional groups, hydrophilicity, mechanical rigidity, recyclability, and simple preparation in varying geometrical configurations, are indispensable to any material deliberated for encapsulating enzymes. Understandably, the non-toxicity and biocompatibility of the materials are also obligatory for food, pharmaceutical, medical, and agricultural applications. Moreover, keeping in view the growing public health and environmental concerns, the materials should also be biodegradable, economical, as well as low-priced [36]. Table 2.1 summarizes the utilization of alginate and chitosan biopolymers as support material in immobilization engineering for various enzymes.

Among the support carriers ascertained for encapsulating enzymes to date, organic or inorganic, natural or synthetic, chitin and chitosan are of particular interest owing to their unique characteristics as stated above. Inarguably, the past decade has witnessed the evidently increasing research investigation on chitin- and chitosan-based materials and a great number of potential bio-based products have developed for applications in industrial arena such as wastewater remediation (elimination of heavy metals, flocculation/coagulation of dyes and proteins, membrane purification processes), the food industry (anti-cholesterol and fat binding, preservative, packaging material, animal feed additive), agriculture (seed and fertilizer coating, controlled agrochemical release), pulp and paper industry (surface treatment, photographic paper), cosmetics and toiletries (moisturizer, body creams, bath lotion) [36, 52–54]. Owing to the unique and characteristics mentioned above, the most important exploitations of chitin/chitosan-based materials, at present, are in the biotechnology sector at large and pharmaceutical/cosmeceutical in particular. In biomedical sector, chitosan or chitosan-based materials are being extensively used as bacteriostatic and fungistatic agents, drug delivery vehicles, drug controlled release systems, artificial cells, wound healing ointments or dressings, hemodialysis membranes, contact lenses, artificial skin, and surgical sutures and for tissue engineering. In biotechnology, they may find application as chromatographic matrices, membranes for separations and notably as enzyme/cell immobilization support [55, 56].

Table 2.1 Applications of alginate and chitosan biopolymers as support material for enzyme immobilization

Enzyme source	Enzyme name	Immobilization support	Immobilization technique	References
Penicillium notatum	Exo-PG	Sodium-alginate matrix	Covalent bonding	[37]
Trametes versicolor	Lac	Chitosan beads	Cross-linking	[38]
Aspergillus terreus	Inulinase	Alginate-gel	Covalent bonding	[39]
Horseradish roots	HRP	Ca-alginate beads	Cross-linking	[40]
Horseradish roots	HRP	Chitosan beads	Entrapment	[41]
Ganoderma lucidum	LiP, MnP, Lac	Alginate-chitosan beads	Cross-linking	[42]
Pseudomonas fluorescens	Lipase	Chitosan	—	[43]
Bacillus licheniformis	Xylanase	Ca-alginate beads	Entrapment	[44]
Trichoderma reesei	Cellulase	Chitosan-coated magnetic nanoparticles	Covalent linkages	[45]
Candida antarctica	Lipase B	Activated chitosan	Cross-linking	[46]
Kluyveromyces marxianus	Inulinase	Chitosan beads	Cross-linking	[47]

(*Continued*)

Table 2.1 (*Continued*)

Enzyme source	Enzyme name	Immobilization support	Immobilization technique	References
Horseradish roots	HRP	Ca-alginate beads	Cross-linking	[48]
Leuconostoc mesenteroides B-512F	Dextransucrase	Chitosan	Covalent bonding	[49]
Coriolopsis gallica	Lac	Chitosan nanoparticles	Encapsulation	[50]
Bacillus licheniformis	Polygalacturonase	Ca-alginate beads	Entrapment	[51]

Chitin- and chitosan-based materials are employed in most of the applications, in the form of powders, flakes, and gels of different geometrical configurations as enzyme immobilization supports. Chitin/chitosan powders and flakes are commercially available (Sigma-Aldrich or Fuji Spinning Co. Ltd. Tokyo, Japan) or can be manufactured in a laboratory. Chitosan gels in the form of beads, capsules, coatings, membranes, fibers, and sponges can be developed. Different follow-up modifications could be utilized to enhance gel stability and robustness characteristics [36]. It is well-known that chitosan is a noteworthy base material for encapsulating a large variety of protein/enzymes since it displays enhanced thermos-stability and recyclability characteristics compared to the natural, free counterpart. Urease has been reported to covalently immobilize onto glutaraldehyde activated chitosan membrane, to provide resistance to the influence of inhibitors, such as boric acid, thioglycolic acid, sodium fluoride and acetohydroxamic acid [57]. Similarly, resistance to the mechanical stirring of D-amino acid oxidase (a flavoprotein using FAD as a cofactor) has been provided by enzyme immobilization on cross-linked chitosan matrix.

2.5 Biopolymers and Environmental Engineering

Evidently, environmental contaminants are posing serious health-related threats. Among potential environmental pollutants, phenolic compounds, polycyclic aromatic hydrocarbons (PAHs), and organic pesticides and herbicides have been considered more critically, and much of literature is available on their potentially toxic effects coupled with poor biodegradability. Several researchers have proposed a bundle of approaches, and each of them has its advantages and disadvantages. In recent years, several materials have been investigated as a potential alternative as compared to many in practice physicochemical methodologies for wastewater treatment purposes. Chitosan is considered as one of the most valuable bio-based polymeric materials for the adsorption of environmental contaminants. The manifestation of amino and hydroxyl groups in chitosan endorses its adsorption interactions with the above-mentioned environmental contaminants including industrial synthetic dyes, metals and organic complexes [58, 59]. Moreover, the amino and hydroxyl functional groups offer great versatility with cross-linking and grafting type interactions. Such unique interactional linkages, thus, in turn, enhance the absorption efficacy and specificity [60]. For instance, at low pH, cross-linking the functional groups of chitosan enhances the adsorption capacity of chitosan. Likewise, grafting with nitrogen or sulfur ameliorates the specificity and capacity of chitosan for some heavy metal ions [61]. The dye adsorption capability by unmodified chitosan is excellent; however, its low stability has provoked many researchers and environmental engineers to work on modifying them. Different modifications strategies (grafted alkyl, amino, carboxyl, sulfur group; cross-linking, ethylene glycol diglycidyl ether, glutaraldehyde, and tripolyphosphate) were envisioned and attempted to upgrade the mechanical and physical properties in combination with adsorption efficiency. Following these manipulations in chitosan structure, the inherent properties of chitosan were improved acting as a more suitable candidate for the adsorption of different dye pollutants [58]. Chitosan can function as a chelating agent for binding toxic heavy metals by the amine groups in the near-neutral environment. Chitosan engineered with diverse derivatives offers a broad variety of new/modified characteristics for adsorbing particular metal ions [59]. Chitosan-based adsorption of organic pollutants offers high adsorption attitudes, insensitivity to toxic substances, right modifiability as well as repeatability. There is no single mechanism or simple theory to elucidate adsorption characteristics. Some interactions related to the adsorption mechanism include partition, diffusion, cation exchange, hydrogen bond, electrostatic interaction, Van der Waals force and dipole–dipole interactions. Chitosan coupled with polyphenol oxidase (PPO) was employed for eliminating endocrine-disrupting bisphenol derivatives from aqueous solutions based on adsorption of enzymatically generated quinine derivatives on chitosan. The optimum pH and temperature were 7.0 and 40°C for bisphenols. The bisphenol elimination time could be diminished by the addition of more chitosan beads [62].

2.6 Biopolymers and Biomedical Engineering

2.6.1 Chitosan-Based Applications

Wound healing is a process for promoting rapid dermal regeneration and accelerated wound healing. As stated above, the chitosan and its derivatives are biodegradable and biocompatible, as well as exhibit antimicrobial activity and low immunogenicity, which makes them valuable biomaterials for wound healing [30, 63, 64]. It is demonstrated that chitosan provides a three-dimensional tissue growth matrix, trigger macrophage activity, and activate cell proliferation. One role of chitosan worth mentioning is the activity preferment of fibroblasts, macrophages, and polymorphonuclear leukocytes, which in turn enhances granulation and organizes the repaired tissues. It also exhibits a role in accelerating the healing progress along with inhibiting scar formation [65, 66]. In recent years, nanofibrous and adhesive-based chitosan have been fabricated as wound dressing biomaterials [67]. The chitosan adhesive displays robust sticking strength with lacking the sutures or staples requirements thus typically applied during surgery to enhance wound healing. Eminently, it can effectively cease bleeding from blood vessels together with air leakage from the lungs [68].

Several histopathological studies have documented the promising results utilizing chitosan as wound dressing materials in patients undergoing plastic surgery, endoscopic sinus surgery and skin grafting [69, 70]. At presently, a large quantity of chitosan-based wound healing materials is available in the market. Mechanistically, the cationic chitosan attracts the negatively charged red blood cells and platelets on the puncture site and consequently shortening the clot formation in addition to hemostasis time. Furthermore, the chitosan-based biomaterial absorbs wound exudates and then swells up producing a soft gel mass that ultimately increases wound healing [71, 72]. Two chitosan films Chit-LA and ChitAA presented excellent wound-healing efficiency causing complete wound cessation with no scar formation. As an effective wound healer, the chitosan–cellulose composite may protect wounds from extreme dehydration and infection [73].

2.6.2 Alginate-Based Applications

From a pharmaceutical perspective, alginate-based wound dressing offers enormous advantageous for the treatment of acute and chronic wounds. Traditionally used wound dressings (e.g., gauze) contributed a barrier function while keeping the wound dry by allowing evaporation of wound fluids [74]. In distinction, modern alginate-based dressings expedite wound healing by providing a moist wound environment [75]. Indeed, alginate dressings are characteristically produced by ionic cross-linking of an alginate solution with calcium ions to develop a gel, followed by further processing to form freeze-dried porous sheets and fibrous non-woven dressings. In the dry form, alginate dressings absorb wound exudates to re-gel, which in turn can supply water to a dry wound, keeping a physiologically moist microenvironment and consequently curtailing bacterial infection at the wound site.

To date, a great variety of functional and bioactive alginate-based wound dressings have been examined and used for wound healing purposes. Most commonly available commercial alginate dressings include Algicell™ (Derma Sciences) AlgiSite M™ (Smith & Nephew), Comfeel Plus™ (Coloplast), Kaltostat™ (ConvaTec), Sorbsan™ (UDL Laboratories), and Tegagen™ (3M Healthcare). Outstandingly, alginate gels releasing stromal cell derived factor-1 were found to be effective in enhancing wound healing rates and decreasing scar formation [76]. Integration of silver into alginate dressings accelerated the antimicrobial and antioxidant capability along with improving the binding affinity for elastase and pro-inflammatory cytokines [77]. The zinc ions cross-linked alginate fibers have shown potential for wound dressings, since zinc ions may generate antimicrobial immune-modulatory effects, as well as enhanced levels of endogenous growth factors [78]. Composites of alginate, chitin/chitosan, and fucoidan gels have also been experimented to provide a moist healing microenvironment for simple application and removal [79, 80].

2.7 Final Remarks and Future Trends

In conclusion, biopolymers are approaching the reality of substituting conventional synthetic counterparts. Without any claims to broadness, the current scientific literature summarizes that bio-based polymers have found their place in various commodities and state-of-the-art industrial applications due to the main advances in biotechnologies and public consciousness. Despite great promise, there are still some downsides impeding their commercialization mainly due to performance and price when compared with their pristine existing counterparts. A comprehensive overview of alginate and chitosan-based biopolymer implies that there continues to be vivid interest in exploiting bio-based materials, especially since a promising enzyme immobilization support for myriads of biotechnological applications ranging from removing organic pollutants from wastewaters to sophisticated hi-tech biosensors development. In conclusion, these kinds of studies could play a pivotal role in advancing the hitherto underutilized, renewable biopolymer of high potential to the market of biomaterials.

Challenges that need to be circumventing in the upcoming years include efficient management of raw materials, improving the functionality of bio-based materials, and their cost for production. The economy of scale would be one of the foremost challenging tasks for the manufacture of bio-monomers and biopolymers from natural resources. Development of industrial-scale plants can be challenging due to the dearth of experience in novel technologies and assessment of supply/demand equilibrium [11]. Given exhaustive efforts, the current bio-based industry is introducing new bio-based polymers with greater performance and value as well as higher thermal and mechanical properties. Many developments are presently in progress to develop various bio-based polymers with a high distinction in their final properties for utilization in automotive, electronics, and biomedical applications. The use of nanoparticles as additives has long been recognized for petroleum-based polymers. To date, several nano-supports being fabricated include carbon nanotubes (CNTs), graphene, nano-clays, 2-D layered materials, and cellulose nanowhiskers. Uniting these nano-reinforcements with bio-based polymers might augment a large number of physiochemical properties, including flame resistance, thermal stability, solvent uptake, and biodegradability, comparative to the native polymer. Advancements such as nano-reinforcement coupled with bio-based polymers are a very attractive direction to develop new functional biomaterials for diverse applications.

Acknowledgements

The authors are grateful to the Shanghai Jiao Tong University, Shanghai 200240, China, and Tecnologico de Monterrey, Mexico, for providing literature facilities.

Conflict of Interest

The authors do not have any conflicting, competing, and financial interests in any capacity.

References

1. Iqbal, H. M. N. (2015). *Development of Bio-Composites with Novel Characteristics through Enzymatic Grafting.* Doctoral dissertation, University of Westminster, London, UK.
2. Bedian, L., Rodríguez, A. M. V., Vargas, G. H., Parra-Saldivar, R., and Iqbal, H. M. N. (2017). Bio-based materials with novel characteristics for tissue engineering applications—A review. *International Journal of Biological Macromolecules*, 98, 837–846.
3. Iqbal, H. M. N., Kyazze, G., Tron, T., and Keshavarz, T. (2014). "One-pot" synthesis and characterisation of novel P (3HB)–ethyl cellulose based graft composites through lipase catalysed esterification. *Polymer Chemistry*, 5(24), 7004–7012.
4. Iqbal, H. M. N., Kyazze, G., Tron, T., and Keshavarz, T. (2014). Laccase-assisted grafting of poly (3-hydroxybutyrate) onto the

bacterial cellulose as backbone polymer: Development and characterisation. *Carbohydrate Polymers*, 113, 131–137.

5. Iqbal, H. M., Kyazze, G., Tron, T., and Keshavarz, T. (2014). A preliminary study on the development and characterisation of enzymatically grafted P (3HB)-ethyl cellulose based novel composites. *Cellulose*, 21(5), 3613–3621.
6. Iqbal, H. M. N., Kyazze, G., Locke, I. C., Tron, T., and Keshavarz, T. (2015). Poly (3-hydroxybutyrate)-ethyl cellulose based bio-composites with novel characteristics for infection free wound healing application. *International Journal of Biological Macromolecules*, 81, 552–559.
7. Iqbal, H. M. N., Kyazze, G., Locke, I. C., Tron, T., and Keshavarz, T. (2015). Development of bio-composites with novel characteristics: Evaluation of phenol-induced antibacterial, biocompatible and biodegradable behaviours. *Carbohydrate Polymers*, 131, 197–207.
8. Iqbal, H. M. N., Kyazze, G., Locke, I. C., Tron, T., and Keshavarz, T. (2015). Development of novel antibacterial active, HaCaT biocompatible and biodegradable CA-gP (3HB)-EC biocomposites with caffeic acid as a functional entity. *eXPRESS Polymer Letters*, 9, 764–772.
9. Iqbal, H. M. N., Kyazze, G., Locke, I. C., Tron, T., and Keshavarz, T. (2015). In situ development of self-defensive antibacterial biomaterials: Phenol-g-keratin-EC based bio-composites with characteristics for biomedical applications. *Green Chemistry*, 17(7), 3858–3869.
10. Iqbal, H. M. N., Kyazze, G., Tron, T., and Keshavarz, T. (2015). Laccase-assisted approach to graft multifunctional materials of interest: Keratin-EC based novel composites and their characterisation. *Macromolecular Materials and Engineering*, 300(7), 712–720.
11. Babu, R. P., O'connor, K., and Seeram, R. (2013). Current progress on bio-based polymers and their future trends. *Progress in Biomaterials*, 2(1), 8.
12. Arevalo-Gallegos, A., Ahmad, Z., Asgher, M., Parra-Saldivar, R., and Iqbal, H. M. N. (2017). Lignocellulose: A sustainable material to produce value-added products with a zero waste approach—A review. *International Journal of Biological Macromolecules*, 99, 308–318.
13. Bilal, M., Asgher, M., Iqbal, H. M. N., Hu, H., and Zhang, X. (2017). Biotransformation of lignocellulosic materials into value-added products– A review. *International Journal of Biological Macromolecules*, 98, 447–458.

14. Asgher, M., Ahmad, Z., and Iqbal, H. M. N (2017). Bacterial cellulose-assisted de-lignified wheat straw-PVA based bio-composites with novel characteristics. *Carbohydrate Polymers*, 161, 244–252.
15. Ahmad, Z., Bajwa, M. A., and Iqbal, H. M. N. (2017). Enzyme-treated Wheat Straw-based PVOH bio-composites: Development and characterization. *BioResources*, 12(2), 2830–2845.
16. Iqbal, H. M. N., Dhama, K., Munjal, A., Khandia, R., Karthik, K., Dadar, M., Sachan, S., Latheef, S. K., and Joshi, S. K. (2017). Tissue engineering and regenerative medicine potentialities of materials–based novel constructs: A review. *Current Regenerative Medicine,* 7, In-Press, DOI: 10.2174/2468424407666170210144332.
17. Lee, K. Y., and Mooney, D. J. (2012). Alginate: Properties and biomedical applications. *Progress in Polymer Science*, 37(1), 106–126.
18. Smidsrød, O., and Skja, G. (1990). Alginate as immobilization matrix for cells. *Trends in Biotechnology*, 8, 71–78.
19. Rinaudo, M. (2008). Main properties and current applications of some polysaccharides as biomaterials. *Polymer International*, 57(3), 397–430.
20. Qin, Y. (2008). Alginate fibres: An overview of the production processes and applications in wound management. *Polymer International*, 57(2), 171–180.
21. Hay, I. D., Ur Rehman, Z., Ghafoor, A., and Rehm, B. H. (2010). Bacterial biosynthesis of alginates. *Journal of Chemical Technology and Biotechnology*, 85(6), 752–759.
22. Albuquerque, P., Coelho, L. C., Teixeira, J. A., and Carneiro-da-Cunha, M. G. (2016). Approaches in biotechnological applications of natural polymers. *AIMS Molecular Science*, 3(3), 386–425.
23. LeRoux, M. A., Guilak, F., and Setton, L. A. (1999). Compressive and shear properties of alginate gel: Effects of sodium ions and alginate concentration. *Journal of Biomedical Materials Research*, 47(1), 46–53.
24. Kong, H. J., Smith, M. K., and Mooney, D. J. (2003). Designing alginate hydrogels to maintain viability of immobilized cells. *Biomaterials*, 24(22), 4023–4029.
25. Kong, H. J., Lee, K. Y., and Mooney, D. J. (2002). Decoupling the dependence of rheological/mechanical properties of hydrogels from solids concentration. *Polymer*, 43(23), 6239–6246.

26. Orive, G., Ponce, S., Hernandez, R. M., Gascon, A. R., Igartua, M., and Pedraz, J. L. (2002). Biocompatibility of microcapsules for cell immobilization elaborated with different type of alginates. *Biomaterials*, 23(18), 3825–3831.
27. Lee, J., and Lee, K. Y. (2009). Local and sustained vascular endothelial growth factor delivery for angiogenesis using an injectable system. *Pharmaceutical Research*, 26(7), 1739–1744.
28. George, M., and Abraham, T. E. (2006). Polyionic hydrocolloids for the intestinal delivery of protein drugs: Alginate and chitosan—a review. *Journal of Controlled Release*, 114(1), 1–14.
29. Roberts, G. A. F. (1992). Structure of chitin and chitosan, In G. A. E. Roberts (Ed.), *Chitin Chemistry,* Macmillan, London, pp. 85–91.
30. Cheung, R. C. F., Ng, T. B., Wong, J. H., and Chan, W. Y. (2015). Chitosan: An update on potential biomedical and pharmaceutical applications. *Marine Drugs*, 13(8), 5156–5186.
31. Shepherd, R., Reader, S., and Falshaw, A. (1997). Chitosan functional properties. *Glycoconjugate Journal*, 14(4), 535–542.
32. Nwe, N., Chandrkrachang, S., Stevens, W. F., Maw, T., Tan, T. K., Khor, E., and Wong, S. M. (2002). Production of fungal chitosan by solid state and submerged fermentation. *Carbohydrate Polymers*, 49(2), 235–237.
33. Borchard, G. (2001). Chitosans for gene delivery. *Advanced Drug Delivery Reviews*, 52(2), 145–150.
34. Bernkop-Schnürch, A., and Dünnhaupt, S. (2012). Chitosan-based drug delivery systems. *European Journal of Pharmaceutics and Biopharmaceutics*, 81(3), 463–469.
35. van de Velde, F., Lourenco, N. D., Pinheiro, H. M., and Bakker, M. (2002). Carrageenan: A food-grade and biocompatible support for immobilisation techniques. *Advanced Synthesis and Catalysis*, 344(8), 815–836.
36. Krajewska, B. (2004). Application of chitin-and chitosan-based materials for enzyme immobilizations: A review. *Enzyme and Microbial Technology*, 35(2), 126–139.
37. Amin, F., Bhatti, H. N., Bilal, M., and Asgher, M. (2017). Improvement of activity, thermo-stability and fruit juice clarification characteristics of fungal exo-polygalacturonase. *International Journal of Biological Macromolecules*, 95, 974–984.
38. Asgher, M., Noreen, S., and Bilal, M. (2017). Enhancing catalytic functionality of *trametes versicolor* ibl-04 laccase by immobilization

on chitosan microspheres. *Chemical Engineering Research and Design*. 119, 1–11.

39. Awad, G. E., Wehaidy, H. R., El Aty, A. A. A., and Hassan, M. E. (2017). A novel alginate–CMC gel beads for efficient covalent inulinase immobilization. *Colloid and Polymer Science*, 295(3), 495–506.
40. Bilal, M., Iqbal, H. M. N., Hu, H., Wang, W., and Zhang, X. (2017). Development of horseradish peroxidase-based cross-linked enzyme aggregates and their environmental exploitation for bioremediation purposes. *Journal of Environmental Management*, 188, 137–143.
41. Bilal, M., Iqbal, H. M. N., Hu, H., Wang, W., and Zhang, X. (2017). Enhanced bio-catalytic performance and dye degradation potential of chitosan-encapsulated horseradish peroxidase in a packed bed reactor system. *Science of the Total Environment*, 575, 1352–1360.
42. Bilal, M., Asgher, M., Iqbal, H. M. N., Hu, H., and Zhang, X. (2017). Delignification and fruit juice clarification properties of alginate-chitosan-immobilized ligninolytic cocktail. *LWT-Food Science and Technology*, 80, 348–354.
43. Cubides-Roman, D. C., Pérez, V. H., de Castro, H. F., Orrego, C. E., Giraldo, O. H., Silveira, E. G., and David, G. F. (2017). Ethyl esters (biodiesel) production by Pseudomonas fluorescens lipase immobilized on chitosan with magnetic properties in a bioreactor assisted by electromagnetic field. *Fuel*, 196, 481–487.
44. Kumar, S., Haq, I., Prakash, J., and Raj, A. (2017). Improved enzyme properties upon glutaraldehyde cross-linking of alginate entrapped xylanase from Bacillus licheniformis. *International Journal of Biological Macromolecules*, 98, 24–33.
45. Sánchez-Ramírez, J., Martínez-Hernández, J. L., Segura-Ceniceros, P., López, G., Saade, H., Medina-Morales, M. A., Ilyina, A., et al. (2017). Cellulases immobilization on chitosan-coated magnetic nanoparticles: Application for Agave Atrovirens lignocellulosic biomass hydrolysis. *Bioprocess and Biosystems Engineering*, 40(1), 9–22.
46. dos Santos, J. C., Bonazza, H. L., de Matos, L. J., Carneiro, E. A., Barbosa, O., Fernandez-Lafuente, R., Santiago-Aguiar, R. S., et al. (2017). Immobilization of CALB on activated chitosan: Application to enzymatic synthesis in supercritical and near-critical carbon dioxide. *Biotechnology Reports*, 14, 16–26.
47. Singh, R. S., Singh, R. P., and Kennedy, J. F. (2017). Immobilization of yeast inulinase on chitosan beads for the hydrolysis of inulin in a

batch system. *International Journal of Biological Macromolecules*, 95, 87–93.

48. Bilal, M., Iqbal, H. M., Shah, S. Z. H., Hu, H., Wang, W., and Zhang, X. (2016). Horseradish peroxidase-assisted approach to decolorize and detoxify dye pollutants in a packed bed bioreactor. *Journal of Environmental Management*, 183, 836–842.
49. Graebin, N. G., de Andrades, D., Bonin, M. C., Rodrigues, R. C., and Ayub, M. A. (2016). Dextransucrase immobilized on activated-chitosan particles as a novel biocatalyst. *Journal of Molecular Catalysis B: Enzymatic*, In-Press, DOI: 10.1016/j.molcatb.2016.12.007.
50. Koyani, R. D., and Vazquez-Duhalt, R. (2016). Laccase encapsulation in chitosan nanoparticles enhances the protein stability against microbial degradation. *Environmental Science and Pollution Research*, 23(18), 18850–18857.
51. Rehman, H. U., Aman, A., Nawaz, M. A., Karim, A., Ghani, M., Baloch, A. H., and Qader, S. A. U. (2016). Immobilization of pectin depolymerising polygalacturonase using different polymers. *International Journal of Biological Macromolecules*, 82, 127–133.
52. No, H. K., and Meyers, S. P. (2000). Application of chitosan for treatment of wastewaters. In *Reviews of Environmental Contamination and Toxicology*, Springer New York, pp. 1–27.
53. Kurita, K. (2001). Controlled functionalization of the polysaccharide chitin. *Progress in Polymer Science*, 26(9), 1921–1971.
54. Dutta, P. K., Ravikumar, M. N. V., and Dutta, J. (2002). Chitin and chitosan for versatile applications. *Journal of Macromolecular Science, Part C: Polymer Reviews*, 42(3), 307–354.
55. Singla, A. K., and Chawla, M. (2001). Chitosan: Some pharmaceutical and biological aspects-an update. *Journal of Pharmacy and Pharmacology*, 53(8), 1047–1067.
56. Khor, E. (2002). Chitin: A biomaterial in waiting. *Current Opinion in Solid State and Materials Science*, 6(4), 313–317.
57. Tiwari, A. K. (2004). Antioxidants: New-generation therapeutic base for treatment of polygenic disorders. *Current Science-Bangalore*, 86(8), 1092–1102.
58. Vakili, M., Rafatullah, M., Salamatinia, B., Abdullah, A. Z., Ibrahim, M. H., Tan, K. B., Amouzgar, P., et al. (2014). Application of chitosan and its derivatives as adsorbents for dye removal from water and wastewater: A review. *Carbohydrate Polymers*, 113, 115–130.

59. Boamah, P. O., Huang, Y., Hua, M., Zhang, Q., Wu, J., Onumah, J., Boamah, P. O., et al. (2015). Sorption of heavy metal ions onto carboxylate chitosan derivatives—a mini-review. *Ecotoxicology and Environmental Safety*, 116, 113–120.

60. Kyzas, G. Z., and Bikiaris, D. N. (2015). Recent modifications of chitosan for adsorption applications: A critical and systematic review. *Marine Drugs*, 13(1), 312–337.

61. Yong, S. K., Shrivastava, M., Srivastava, P., Kunhikrishnan, A., and Bolan, N. (2015). Environmental applications of chitosan and its derivatives. In *Reviews of Environmental Contamination and Toxicology Volume 233*, Springer International Publishing, pp. 1–43.

62. Kimura, Y., Takahashi, A., Kashiwada, A., and Yamada, K. (2015). Removal of bisphenol derivatives through quinone oxidation by polyphenol oxidase and subsequent quinone adsorption on chitosan in the heterogeneous system. *Environmental Technology*, 36(18), 2265–2277.

63. Rhoades, J., and Roller, S. (2000). Antimicrobial actions of degraded and native chitosan against spoilage organisms in laboratory media and foods. *Applied and Environmental Microbiology*, 66(1), 80–86.

64. Iqbal, H. M. N., Villalba, A., Khandia, R., Munjal, A., and Dhama, K. (2016). Recent trends in nanotechnology-based drugs and formulations for targeted therapeutic delivery. *Recent Patents on Inflammation & Allergy Drug Discovery*, DOI: 10.2174/1872213X106 66161213162823.

65. Muzzarelli, R. A., Mattioli-Belmonte, M., Pugnaloni, A., and Biagini, G. (1998). Biochemistry, histology and clinical uses of chitins and chitosans in wound healing. *Exs*, 87, 251–264.

66. Ueno, H., Mori, T., and Fujinaga, T. (2001). Topical formulations and wound healing applications of chitosan. *Advanced Drug Delivery Reviews*, 52(2), 105–115.

67. Azuma, K., Izumi, R., Osaki, T., Ifuku, S., Morimoto, M., Saimoto, H., Okamoto, Y., et al. (2015). Chitin, chitosan, and its derivatives for wound healing: Old and new materials. *Journal of Functional Biomaterials*, 6(1), 104–142.

68. Ishihara, M., Obara, K., Nakamura, S., Fujita, M., Masuoka, K., Kanatani, Y., Maehara, T., et al. (2006). Chitosan hydrogel as a drug delivery carrier to control angiogenesis. *Journal of Artificial Organs*, 9(1), 8–16.

69. Azad, A. K., Sermsintham, N., Chandrkrachang, S., and Stevens, W. F. (2004). Chitosan membrane as a wound-healing dressing: Characterization and clinical application. *Journal of Biomedical Materials Research Part B: Applied Biomaterials*, 69(2), 216–222.

70. Stone, C. A., Wright, H., Devaraj, V. S., Clarke, T., and Powell, R. (2000). Healing at skin graft donor sites dressed with chitosan. *British Journal of Plastic Surgery*, 53(7), 601–606.

71. Nguyen, N., Hasan, S., Caufield, L., Ling, F. S., and Narins, C. R. (2007). Randomized controlled trial of topical hemostasis pad use for achieving vascular hemostasis following percutaneous coronary intervention. *Catheterization and Cardiovascular Interventions*, 69(6), 801–807.

72. Weng, M. H. (2008). The effect of protective treatment in reducing pressure ulcers for non-invasive ventilation patients. *Intensive and Critical Care Nursing*, 24(5), 295–299.

73. Sahoo, D., Sahoo, S., Mohanty, P., Sasmal, S., and Nayak, P. L. (2009). Chitosan: A new versatile bio-polymer for various applications. *Designed Monomers and Polymers*, 12(5), 377–404.

74. Boateng, J. S., Matthews, K. H., Stevens, H. N., and Eccleston, G. M. (2008). Wound healing dressings and drug delivery systems: A review. *Journal of Pharmaceutical Sciences*, 97(8), 2892–2923.

75. Queen, D., Orsted, H., Sanada, H., and Sussman, G. (2004). A dressing history. *International Wound Journal*, 1(1), 59–77.

76. Rabbany, S. Y., Pastore, J., Yamamoto, M., Miller, T., Rafii, S., Aras, R., and Penn, M. (2010). Continuous delivery of stromal cell-derived factor-1 from alginate scaffolds accelerates wound healing. *Cell Transplantation*, 19(4), 399–408.

77. Wiegand, C., Heinze, T., and Hipler, U. C. (2009). Comparative in vitro study on cytotoxicity, antimicrobial activity, and binding capacity for pathophysiological factors in chronic wounds of alginate and silver-containing alginate. *Wound Repair and Regeneration*, 17(4), 511–521.

78. Ågren, M. S. (1999). Zinc in wound repair. *Archives of Dermatology*, 135(10), 1273–1274.

79. Murakami, K., Aoki, H., Nakamura, S., Nakamura, S. I., Takikawa, M., Hanzawa, M., Sato, Y., et al. (2010). Hydrogel blends of chitin/chitosan, fucoidan and alginate as healing-impaired wound dressings. *Biomaterials*, 31(1), 83–90.

80. Wang, T., Zhu, X. K., Xue, X. T., and Wu, D. Y. (2012). Hydrogel sheets of chitosan, honey and gelatin as burn wound dressings. *Carbohydrate Polymers*, 88(1), 75–83.

Chapter 3

3D Printing of Biopolymers: Trends and Opportunities for Medical Applications

Tomy J. Gutiérrez

Thermoplastic Composite Materials (CoMP) Group,
Institute of Research in Materials Science and Technology (INTEMA),
Faculty of Engineering, National University of Mar del Plata (UNMdP) and National Council of Scientific and Technical Research (CONICET),
Colón 10850, Mar del Plata 7600, Buenos Aires, Argentina

tomy.gutierrez@fi.mdp.edu.ar, tomy_gutierrez@yahoo.es

Overall, three-dimensional (3D) printing is a technology booming for processing and development of unique polymeric materials. In this context, the manufacture of parts layer-by-layer with specific requirements such as biomedical devices and tissue engineering may represent an economical option, offering the flexibility of starting materials. However, there are many outstanding challenges to meet. This chapter provides an overview of the field of 3D-printed biopolymers for medical applications, and future trends and opportunities in this field.

Handbook of Biopolymers: Advances and Multifaceted Applications
Edited by Shakeel Ahmed, Suvardhan Kanchi, and Gopalakrishnan Kumar

ISBN 978-981-4800-17-4 (Hardcover), 978-0-429-02475-7 (eBook)
www.panstanford.com

3.1 Introduction

Since the past decade, tissue engineering has shown a sensational promise in providing more viable alternatives to surgical procedures for harvested tissues, implants, and prostheses. Due to the fast development on biomaterial technologies, it is now possible for doctors to use patients' cells to repair orthopedic defects such as focal articular cartilage lesions. In order to support the three-dimensional tissue formation, scaffolds made by biocompatible and bioresorbable polymers and composite materials, produced by three-dimensional (3D) printing, for providing temporary support to the damaged body and cell structures have been developed recently. Although ceramic and metallic materials have been widely accepted for the development of implants, their non-resorbability and necessity of second surgical operation result in extra discomfort for the patients and limit their wide applications [1].

3D printing is also known as additive manufacturing (AM), layered manufacturing, rapid prototyping (RP), or solid freeform fabrication. This technique represents the direct fabrication of parts layer-by-layer, based on data obtained by computerized medical imaging equipment such as magnetic resonance imaging (MRI) and computed tomography (CT) scans, using a computer-aided design (CAD) file [2]. In 3D printing, CAD models of the parts to be manufactured are first sliced in a virtual environment to create a stack of two-dimensional (2D) slices. A 3D printing machine then builds the parts one layer at a time based on the 2D slice information, stacking and joining successive layers to make the final 3D object. Just as the Internet has given us the ability to access information and connect with people from different parts of the world, CAD has provided us the ability to create, modify, and, if needed, critique designs in a virtual world. With the advent of 3D printing, such virtual designs can now be rendered into physical 3D objects that can serve as prototypes or be directly used as functional parts for a variety of applications. In this context, polymers in various forms, reactive, liquid solutions or as thermoplastic melts play a key role in many applications and the further expansion toward manufacturing robust, real end use products [3].

The origins of contemporary 3D printing can be traced back to the 1980s when Hull invented stereolithography (SLA), the first 3D printing technology [4]. SLA is a process in which an ultraviolet (UV) laser light source is focused onto the surface of a UV-curable liquid monomer bath and scanned in patterns representing slice cross sections. The scanned monomers undergo photo-induced cross-linking and harden to form the desired 2D cross sections, while the uncured monomers remain in the bath. Hull was also the first to find a way to allow a CAD file to communicate with the RP system in order to build computer-modeled parts. Hull's patent was approved in 1986, making it the first patent for a 3D printer. The company 3D Systems, founded by Hull, focused on commercializing SLA systems, which were the first commercial 3D printers [5].

Fast forward to 2014, when the U.S. National Aeronautics and Space Administration (NASA) launched the first 3D printing machine to the International Space Station to directly build parts in space under zero gravity [6]. This is also the year when researchers from Oak Ridge National Laboratory built a complete car body using a 3D printing technique known as big area additive manufacturing and partnered with Local Motors to commission and drive a functional car at the International Manufacturing Technology Show [7]. 3D printing-based agile manufacturing technologies are promoting on-demand production with traditional as well as innovative designs that are difficult, if not impossible, to make through conventional manufacturing (CM) approaches. 3D printing technologies are also making a significant impact in biomedical research—from device designs to tissue engineering (TE) to bioprinting and drug delivery.

After following the technology, markets, and publications over several years, it appears that now we are in a period where many decisive actors in industry, media, and research and even finance advisors realize that the traditional area of prototyping is expanding more and more into manufacturing, and this will have a significant impact on many areas of our lives, and probably a spectacular growth of the 3D printing industry [3].

3.2 Brief History of 3D Printing

Several 3D printing technologies were conceived and developed around the time of the emergence of SLA. Deckard invented selective laser sintering (SLS) as a graduate student in Beaman's group at the University of Texas at Austin [8]. SLS uses powder materials spread on a build plate where a laser selectively sinters the powder in certain areas based on the CAD file [6]. A similar powder bed-based concept formed the basis of another technology, 3D printing, at the Massachusetts Institute of Technology by Sachs' group. Inkjet printing was combined with a powder bed, where a binder was printed onto each successive layer of swept powder based on the CAD slice information. Using this approach, complex shaped metal, polymer, and ceramic parts could be printed. However, post-processing or sintering steps were often required to enhance the final strength of the parts [9]. Scott and Lisa Crump developed another 3D printing technology called fused deposition modeling (FDM). FDM involves heating a thermoplastic filament to a semi-liquid state, which is then extruded and deposited through a nozzle onto a substrate to build parts layer-by-layer based on the CAD file information [10, 11]. Additionally, Sanders released the first 3D printer based on inkjet printing of thermoplastic polymers [12]. Parts with fine features could be made easily using this approach. These are some of the notable early-stage 3D printing technologies that were primarily focused on RP for design verification and visualization.

Over the past 15 years, a variety of new technologies have evolved that have transformed the idea of RP to AM, where parts produced by a 3D printer can be directly used for a variety of applications. In the case of metallic materials, laser-based or electron beam-based technologies with or without a powder bed have truly revolutionized industrial applications of these printers. For biomedical applications, novel fabrication approaches based on the 3D-Bioplotter or direct ink writing, laser-assisted bioprinting, and robotic-assisted printing are all in use for different applications. Table 3.1 offers a brief summary of some of these technologies that are relevant for biomaterials and their applications toward TE [13–26]. Other technologies related to metallic printing are covered throughout this chapter.

Table 3.1 3D printing technologies and their applications in biomaterials

Technique	Process details	Processed materials for tissue engineering	Advantages and disadvantages	References
3D plotting/direct ink writing	Extrusion-based layer-by-layer deposition	A variety of polymers and ceramics have already been used, including polycaprolactone (PCL), hydroxyapatite (HA), bioactive glasses, polylactic acid (PLA)/polyethylene glycol (PEG), and poly(hydroxymethyl glycolideco-ε-caprolactone)	**Advantage** –Easy to incorporate both drug and biomolecules (proteins and living cells) **Disadvantages** –Post-processing may be needed for some materials –Process works well within a certain range of viscosity	[13–16]
Laser-assisted bioprinting	Coating the desired material on transparent quartz disk (ribbon); the deposition is controlled by laser pulse energy, and resolution is controlled by distance between ribbon/substrate, spot size, and stage movement	A variety of materials, including HA, zirconia, HA/MG63 (osteoblast-like cell), human osteoprogenitor cell (i.e., a cell that has the potential to transform into one that forms bone), and human umbilical vein endothelial cell	**Advantage** –Ambient and mild condition is suitable for organic and inorganic materials and cells **Disadvantage** –Homogeneous ribbons are needed	[17, 18]

(*Continued*)

Table 3.1 (*Continued*)

Technique	Process details	Processed materials for tissue engineering	Advantages and disadvantages	References
Selective laser sintering	A powder bed-based process that partially or completely sinters layer-by-layer using a laser-based heat source	A large variety of polymers, metals, and ceramic materials have been used with this technique, including PCL, HA, PLLA, tricalcium phosphate, and poly(3-hydroxybutyrate)	**Advantages** –Powder bed is used as support, therefore no need for secondary support structures –Can be used for a variety of materials **Disadvantage** –Feature resolution depends on laser-beam diameter	[19, 20]
Stereolithography	Layer-by-layer fabrication by exposure to photopolymer liquid; polymer solidifying at the focal point, and un-exposed polymer remains liquid	A large variety of photo-curable polymers	**Advantage** –Simple and complex designs can be manufactured; growth factors, proteins, and cells can also be done **Disadvantage** –Only applicable for photopolymers	[21, 22]

Technique	Process details	Processed materials for tissue engineering	Advantages and disadvantages	References
Fused deposition modeling	Strands of thermoplastic polymers or polymer/ceramic composites extruded through a tip and deposited layer-by-layer	Structural and biopolymers, ceramic-polymer, or metalpolymer composites	**Advantage** –Easy to use and can be used with a large variety of materials **Disadvantage** –Material restriction related to thermoplastic polymers	[23–25]
Robotic assisted deposition/robocasting	Direct writing of high solids loaded slurry; good for a variety of ceramics and ceramic-polymer composites	HA/PLA, HA/PCL, and bioactive glass (6P53B)/PCL	**Advantage** –Good for ceramics **Disadvantage** –May not be useful for different materials	[26]

An American Society for Testing and Materials (ASTM) International committee dedicated to the specification of standards for AM was formed in 2009 [27]. This committee, known as ASTM F42, created a categorization of all 3D printing technologies into seven major groups. Table 3.2 shows the major categories along with the well-known 3D printing technologies that fit within each category.

Table 3.2 3D printing techniques and respective vendors

Category	Commercial 3D printing technologies and vendors
Vat photopolymerization	• Stereolithography from 3D Systems • Bioplotters from Envisiontec • Large area maskless photopolymerization from DDM Systems • Lithoz Lithography-Based ceramic manufacturing
Material extrusion	• Fused deposition modeling from Stratasys
Powder bed fusion	• Selective laser sintering from 3D Systems • Electron beam melting from Arcam AB • Direct metal laser sintering from EOS • Selective laser melting from SLM Solutions
Directed energy deposition	• Laser engineered net shaping from Optomec Inc • Direct metal deposition from DM3D • Electron beam welding from Sciaky Inc
Material jetting	• Objet from Stratasys • Solidscape 3D printers from Solidscape • Multi-jet Fusion Technology from HP
Binder jetting	• ZCorp • ExOne • Voxeljet
Sheet lamination	• MCor Technologies

3.3 3D Materials Processing Techniques

In general, there are two modes of processing 3D materials: the extrusion mode and the droplet mode. In the extrusion mode, the material is extruded out of the nozzle tip under an applied pressure. This mode can basically lay down the material in the form of line structures to create the desired model by moving the nozzle tip over a substrate in the designed path. This process can be repeated layer by layer to develop a freeform fabricated part. In the droplet mode, the material is deposited in the form of droplets that is controlled by using a frequency function and key parameters in the nozzle system settings. The droplet mode can form a structured layer by depositing multiple droplets at desired locations on a substrate. Similarly, this process can be repeated to fabricate a 3D structure. Figure 3.1 shows a schematic diagram of the extrusion and droplet mode for the deposition.

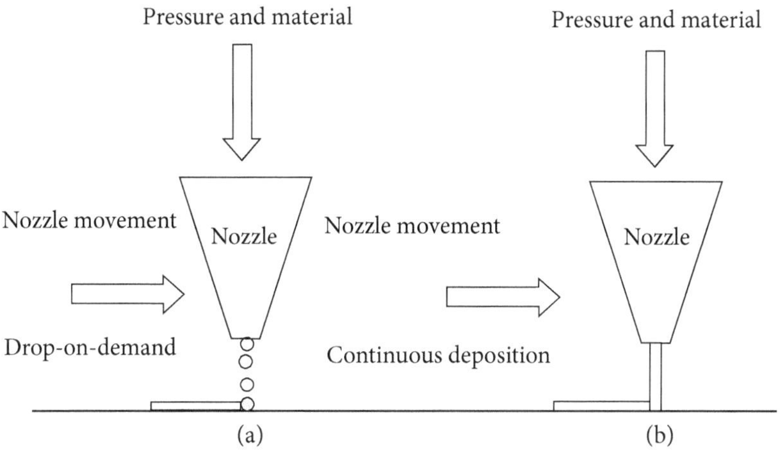

Figure 3.1 3D materials processing techniques: (a) extrusion mode and (b) droplet mode.

Each nozzle system is unique in its method of operation, which makes each system have its advantages and limitations over the others. All nozzle systems have a material delivery system. However, the detailed setup for each system is different and the material delivery system parameters such as air pressure are not controlled by in-house software. Characteristics and comparison of the four nozzle systems with their advantages and limitations is shown in Table 3.3.

Table 3.3 Characteristics and comparison of the four nozzle systems

	Microvalve nozzle system			
Features	**Pressurized mini extruder**	**Solenoid micro-nozzle**	**Piezoelectric micro-nozzle**	**Pneumatic micro-nozzle**
Deposition mode	Continuous	Continuous/droplet	Droplet	Continuous/droplet
Operation/control	Rotating screw gear via motor	Frequency pulse of voltage	Frequency pulse of voltage	Frequency pulse of air pressure
Key process parameters	Pressure and speed, temperature, material, nozzle diameter, deposition speed	Pressure, frequency pulse, material, nozzle diameter, deposition speed	Pressure, frequency pulse, material, nozzle diameter, deposition speed	Pressure, frequency pulse, material, nozzle diameter, deposition speed
Operating range limitations	Screw speed: 1 rps Temperature: 150°C *D*: 7–10 mil	*V*: 40 V(DC) *H*: (1–1,200 Hz) *D*: (30, 50, 70 μm)	*H*: (0–20,000 Hz) *V*: (2100-100) *D*: (30, 50, 70 μm)	*H*: (0.01–14 Hz) Fluid *P*: (0–50 psi) Valve *P*: (70–100 psi)
Structure formation	Physical solidification	Physical solidification, chemical reaction	Physical solidification, chemical reaction	Physical solidification, chemical reaction
Advantages	Fast solidification, no solvents, strong structure, sterile environment	Room temperature, extrusion and droplet, sterile environment	Room temperature, micro-droplet deposition, controlled volume, sterile environment	Room temperature, high viscosity, extrusion and droplet, sterile environment
Disadvantages	Temperature, low melting material	Low viscosity, droplet controllability	Low viscosity, not continuous deposit	Droplet controllability, precision deposition

More specifically, these methods of processing 3D materials can be divided into two distinct groups based on the step, which defines the shape of each cross section.

As a summary, the first group can be described as a repetitive process of the following:

(i) Apply a "fluid" material as a thin layer on a workpiece (or fluid bed) under construction.
(ii) Induce a selective phase change (solidification) on the current cross section of the work piece.
(iii) Lower the work piece (on a platform) into the fluid bed by one layer thickness; then repeat.

The second group combines the "imaging" step with the addition of the material of a full cross section:

(i) Dispense a liquid or liquefied material as small droplets or filament onto the preceding cross section of the work piece.
(ii) Solidify the new cross section.
(iii) Optionally equalize the new layer to a defined thickness; then repeat.

A third group is the group of "imaging" on a bulk material that uses a (ink) jet technology on a powder bed (ASTM: Binder Jetting) to solidify the loose powder with a liquid solution, which dries or reacts to agglomerate the powder particles at the specified locations of each layer. This method was originally coined with the name "3D printing" by the MIT, but is now used alongside the other techniques with the same designation.

If we look at the second group of technologies where we use addition of the material for each full cross section, without a separate imaging step, we have two major processes, which dominate the market by machine numbers: (Multi)jet printing of reactive resins or molten wax (ASTM: Material Jetting) and extrusion of molten thermoplastics (ASTM: Material Extrusion).

The processes coined as Multijet Modelling or Polyjet use a print head similar to inkjet printers to deposit exactly shaped layers of a photopolymer resin, and normally a support material to partially surround it, which is then immediately cured with a UV lamp. In order to maintain exact layer thickness and compensate for small variations of droplet size, current machines use a mechanical planarizer, which removes excess material on each pass of the print head.

These machines can process robust resin objects—generally complex acrylate formulations—with high accuracy and resolution, and do not require complex lasers and beam scanners. Some limitations related to materials relate to the viscosity requirements of the jetting process; often the viscosity is reduced by heating the print head and nozzle plate. This heating is also required in machines, which are processing wax materials to produce patterns for investment casting, in jewelry and dental restorations applications notably.

A recent development adds machines with multiple print heads or channels, again similar to color inkjet printers, which can deposit selected mixtures of different materials. Besides a palette of colors, this also allows to assign different material properties, essentially modulus of elasticity, to different regions of a part. It is another step to replace assemblies of several components and fasteners with a single part, which is generally an advantage of additive technologies.

The last technique we look at is based on the original process of FDM and also relies on a rather simple buildup of an essentially standard thermoplastic material (ASTM: Material Extrusion). The concept of melting a "standard" plastic material and extruding it through a thin, heated nozzle seems quite straightforward; the extruded melt can then be spread in a controlled X-Y-movement onto the previous layer of part being built. Obviously, the processing temperature has to be adapted to the specific material and well controlled, and thermal shrinkage after deposition can be reduced by heating the build platform or even the complete build chamber.

Originally developed and implemented as an industrial quality process with larger size machines and a selection of thermoplastic polymers, the scope has now grown into the widest palette of smaller and low-cost machines. Some of these machines are even offered as kits to be built or assembled by the user. Recently, this type of machines have become very popular and also have the largest number manufacturers of quite similar devices, mostly with rather small build volumes of roughly 100 to 200 mm on each side. While the simple machines just have one extrusion head and filament spool, many of the newer developments have multiple heads and can extrude support and part material for good accuracy and complex shapes, or to combine different colors in a single part.

The most popular polymers in this process are ABS and poly lactide, which is also offered on almost every entry-level printer, but higher-grade machines can also process engineering materials such as polycarbonate, polyamide, or even polyphenyl sulfone. As the starting material in this process already is a thermoplastic polymer, the mechanical properties of parts approach those of traditionally molded or machined products, similar to SLS, but using a significantly simpler machine.

The extrusion-type machines are a specific example that demonstrate the accelerated progress in the development toward the wide availability of 3D printers at affordable cost, relying on the assembly of readily available core components such as precision rails, servo drives, control electronics modules, including computers and software that can handle the complex geometry data. The latter is also essential for the design of the 3D model data based on CAD programs or object scanners; not the least is the declining cost of computers to handle all these data and control the machine components.

Another early method that uses paper or other thin films of material, which are stacked up, bonded together, and cut to shape was called layered object manufacturing (LOM) and has now been revived (ASTM: Sheet Lamination) with the possibility to use standard office paper sheets. Whereas the original version used an IR laser to cut the individual layers, the more recent devices cut the paper with a stencil, and a special option allows printing onto the sheets with an office printer, which then creates fully colored parts. Whereas paper is a well-known standard build material, the pressure- or heat-sensitive adhesives which hold the sheets together are specially developed to fit the process.

It has been observed that a vast choice of 3D printing methods is available, many with a variety of materials to choose from, and generally a large range of possible or preferred applications. Technically the possibilities are almost without limits, and on top of the thousands of applications that are being realized today, there is also a lot of speculation about the imminent replacement of current production technologies with these new 3D printers. In reality, the majority of users still apply these methods to produce prototypes of all kinds, unique parts and new designs, including architectural models and display objects. There are quite some areas where 3D printing is competitive with the cost and speed required for a specific set of products, and the available materials meet specifications. Today these applications are mostly in the medical field for personalized prostheses, in the aeronautical business for low-volume production of special parts, or complex-shaped tools and fixtures for traditional manufacturing. The benefits are particularly high for complex parts, which can often replace an assembled device with a single print. The user community is constantly expanding the limits toward manufacturing by 3D printing, and with progress on the machine side, improved reliability, build speed, and intuitive user interfaces, the range of these applications keeps growing steadily. Besides that effort, there is a lot of development in materials to expand the scope of manufacturing, notably by the machine manufacturers, independent materials supply companies and at research institutions.

This quest for better materials is also true in the realm of low-cost printers, which are affordable for individuals and schools at higher levels; hobby designers, artists, and researchers can turn their ideas and creations into solid plastic parts. There is notable interest in creating toys, accessories for personal electronics or sometimes even replacements for broken parts.

3.4 Biopolymers Used for 3D Printing

Different biopolymers have been studied as materials for the development of devices printed in 3D to be used in medical applications, such as chitosan, alginate, and collagen, among others [2, 28]. Other than biopolymers, poly(ethylene glycol) (PEG), poly(ethylene oxide) (PEO), polycaprolactone (PCL), and poly-L-lactide acid (PLLA) are the widely used synthetic polymers in the biomedical field [29].

However, not all polymeric materials are applicable and suitable for all manufacturing methods. For example, molding methods are inappropriate for developing medical implants from hydrogels. Because, the biomaterial cannot be removed without damaging both internal and external architecture, which will most likely cause the implant breakage [29].

Another factor to consider is the degradation rate of the biomaterial, since it is desired that the material degrade once its function is accomplished, which would avoid a second operation on the individuals to remove the implant. In this context, the biodegradation rate of PLLA is too low to match the tissue regeneration process after implantation. Besides, the acidic degradation products of PLLA, such as lactic acid, tend to cause aseptic inflammation in tissue [29]. On the contrary, PCL has the presence of hydrolytic aliphatic–ester bonds, which are unstable. Therefore, its degradation duration can be as long as 24 months for being completely degraded. Thus, PCL is always used to co-polymerize with other materials to have desired degradation properties [1]. So, PCL has a better degradation rate to make long-term implants and controlled-drug-release implants. In this sense, Honda et al. [30] used poly(L-lactide-ε-caprolactone) as a biodegradable sponge and implanted it into a nude mice. After a 4-week experimental process, the result showed that there was the formation of cartilage-like structures in the construct [1].

Additionally, polymers can be chemically modified to match the rate of tissue regeneration and degradation rate of the material as well as to obtain a wide range of properties for biomedical applications, such as mechanical properties, diffusivity, density, and hydrophilicity. In addition, these may have optimal control over specific cellular interactions with the polymer material, whereby cells are comfortable, interacting with proteins that bind to the surface of the material, particularly biopolymers [29].

On the other hand, the combination of more than one biomaterial provides better characteristics than any single type of biomaterial, which in turn leads to the development of composite polymeric materials [29]. In this context, composite polymers as 3D printing materials have been used to improve implant–tissue biocompatibility, i.e., they will not be rejected by the body upon implantation [2]. For example, Inzana et al. [31] developed osseous implants from polymer blends of collagen and Tween 80 using calcium phosphate as a natural filler with the aim of improving osteointegration and improving the mechanical properties of the implant [31]. However, the development of these materials requires high temperatures that prevent the incorporation of bioactive molecules and drugs during the 3D printing process, which could stimulate bone regeneration or combat infections [31].

The feasibility of 3D printing of composite biopolymers at low temperature has also been demonstrated from low-melting-point matrices [2, 31]. This has allowed the incorporation of drugs, cells, and 3D printing of synthetic or biological polymers such as collagen, thus allowing the development of bioactive implants [2]. Type I collagen is the most abundant structural protein in the human body and is a critical component of bone extracellular matrix, where it plays important roles in this mineralized tissue's strength and toughness. It has been shown that incorporating collagen into mineralized bone cements could enhance their biomechanical properties, as well as their osteoconductive and osteoinductive characteristics. For example, collagen incorporation into hand-mixed calcium phosphate cements has been shown to improve cellular attachment, viability, proliferation, and activity as well as mechanical properties. Advantages of 3D printing over molding or paste injection include patient-specific geometries and controlled spatial patterning of drugs or polymers within the scaffold [31].

3.5 Advantages of 3D Printing for Medical Applications

3D printing technology offers significant advantages for medical applications due to the ability to manufacture low-volume or one-of-a-kind parts on-demand based on patient-specific needs. For example, surgical implants are currently manufactured by making a near-net-shape part via forging, casting, or machining operations, followed by specialized surface finishing or treatments for the desired surface, mechanical properties, and aesthetic effects, as shown in Fig. 3.2. These operations require expensive tooling; therefore, patient-specific or one-of-a-kind implants are costly and are rarely used. Other challenges such as difficulty in machining of titanium alloys due to high strength, low modulus of elasticity, and low thermal conductivity compared with steel makes it more expensive to manufacture patient-specific implants from these materials [32]. Finally, CM technologies can be energy intensive, producing significant amounts of material waste, and are not capable of easily producing implants with functional gradation.

3D printing or AM represents a new option for the production of a variety of biomedical devices such as orthopedic implants. Although AM may require final machining, the AM-based approach allows significant flexibility toward manufacturing customized, low-volume, complex implants. Figure 3.2 compares CM of implants with AM. AM provides geometrical freedom to designers without manufacturing constraints, leading to novel lightweight designs and potentially reduced part counts for medical implants. Specifically for medical implants, AM allows for customized complex shaped functional implants and demand-based manufacturing—which can offer a significant reduction in cost and inventory. Since AM does not require any part specific tooling, cost per part remains constant for AM. However, for CM, there is always a fixed cost for tooling and dies, and therefore, cost per part decreases as the volume of parts increases until it reaches a minimum. Such cost analysis forms the basis and rationale for the use of 3D printing or AM for biomedical devices.

Despite some noteworthy success, the construction of human tissue or entire organs with 3D printing or AM continues to present significant challenges [33–36]. From plastic surgery to cancer treatment and from treatment of birth defects to prosthetics for amputees—all areas of medicine are seeking breakthroughs enabled by 3D printing to enhance quality of life or to help patients live longer.

We present two examples that illustrate future possibilities. Complex craniofacial titanium implants from computed tomography images of a fractured skull, shown in Fig. 3.3, can be fabricated directly using a laser-based AM technology, Laser Engineered Net Shaping. In another example, shown in Fig. 3.4, TE scaffolds for bone regeneration can be produced through SLS of poly(caprolactone), a biocompatible and bioresorbable polymer, and poly(caprolactone)-hydroxyapatite composites [37, 38]. Furthermore, the scaffolds can be produced with a priori designed mechanical properties to match the desired mechanical performance of the target bone tissue.

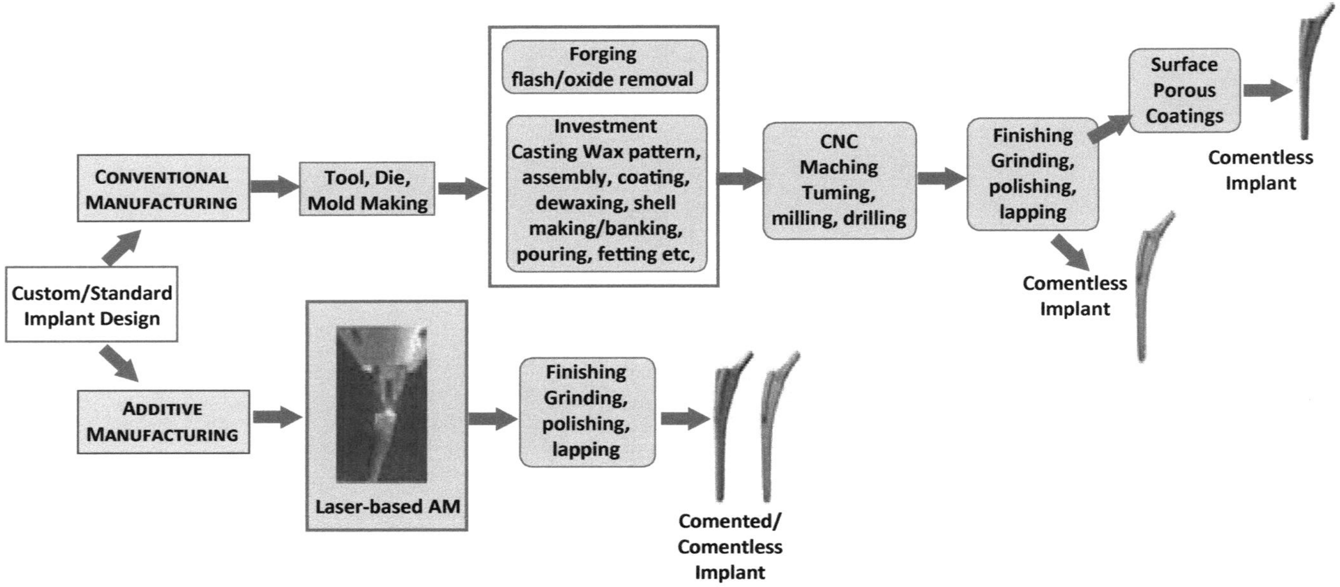

Figure 3.2 Comparison of additive and conventional manufacturing processes of implants.

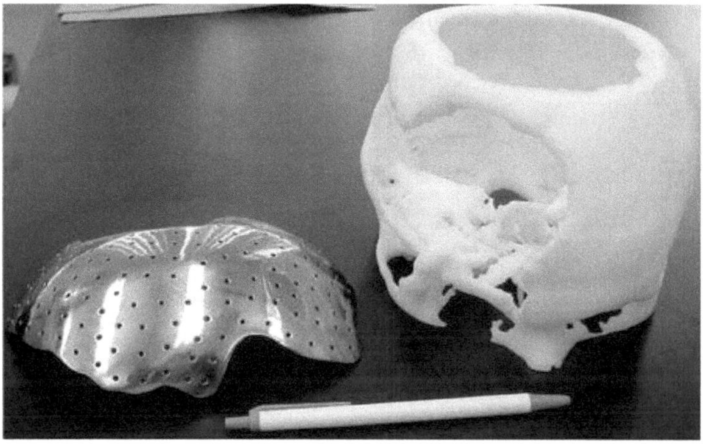

Figure 3.3 Laser Engineered Net Shaping processed craniofacial Ti implant (left) and fused deposition modeling processed polymer prototype of the skull with large defect (right).

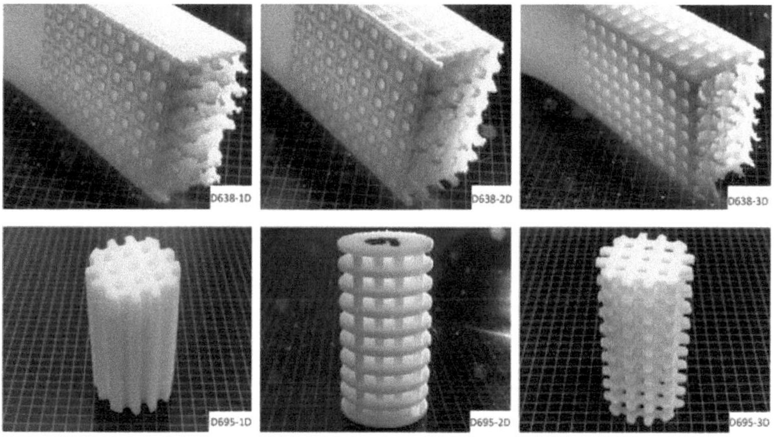

Figure 3.4 Selective laser sintering-processed poly(caprolactone) tensile specimens shown post-fracture (D638-1D, D638-2D, and D638-3D) and compressive specimens (D695-1D, D695-2D, and D695-3D) based on ASTM D638 and D695 test specimen geometries endowed with 1D, 2D, and 3D orthogonal porous channels (placed on a 2 mm grid).

It is envisioned that application of 3D printing will revolutionize the field of biomedical devices and TE due to its inherent flexibility in manufacturing complex parts using various materials. However, many challenges still remain, due to a lack of different materials for use in any printer and the inability to manufacture multicomponent structures using commercial printers.

3.6 Current Challenges in 3D Printing of Biomaterials

3.6.1 Achieving Target Material Properties and Desired Architectures

By virtue of layer-by-layer additive fabrication, it is claimed that 3D printing provides "complexity for free," allowing the physical realization of complex 3D structures endowed with complex internal and external architectures. However, every layer and every interface between adjacent layers provides the opportunity for the introduction of defects during 3D printing that can lead to the failure of the part under the intended application conditions.

Some of the key challenges associated with 3D printing of biomedical implants lie in achieving target mechanical properties, durability, and designed architectures to satisfy both mechanical and functional requirements. This is especially important in load-bearing implants, where mechanical performance and fatigue are critical. 3D printing technologies are being used to produce permanent implants in metals, polymers, and ceramics. The capabilities of these technologies in processing feedstock materials such as titanium, poly(ether ether ketone), and hydroxyapatite need to be well-characterized and documented under diverse conditions both in vitro and in vivo.

The challenges increase when the implants are endowed with intentionally designed or manufacturing-induced porous architectures. The effective mechanical properties and failure modes of such porous architectures need to be well understood and integrated into implant design procedures to avoid undesirable implant failure outcomes. Computational design techniques validated with experimentally measured property data need to be integrated into implant design procedures. Additional challenges lie in the ability of 3D printing technologies to accurately produce desired porous architectures meeting design intent. Often, the length scales associated with the designed porous architectures (ranging from tens to hundreds of microns) are beyond the resolution capability of the 3D printing technology in use.

Finally, surface finish or surface roughness can be a critical requirement as well. The inherent stair-stepped surface finish from layer-by-layer fabrication or other material consolidation-induced surface roughness artifacts prevalent in 3D printing are often beneficial for cell anchoring, proliferation, and integration, but in other cases, they can be highly detrimental due to the inducement of foreign body reactions or due to the release of particles into regions surrounding the implant. Thus, careful attention has to be paid to understanding surface finish and surface texture requirements as well as choosing appropriate post-processing procedures to attain final desired surfaces in biomaterials fabricated by 3D printing.

3.6.2 Clean and Sterile Manufacturing Environments

3D-printed biomaterials are being produced with the intention of use in implants or in laboratory scale studies in bioengineering. In either case, careful attention needs to be paid to the conditions under which the feedstock material is produced, supplied, processed, post-processed, and handled. The scale-up of 3D printing technologies for biomaterials will require careful design of clean manufacturing environments both inside and outside the 3D printing platforms to ensure the avoidance of material contamination and related possible infections. Thus, the design of 3D printing equipment for biomaterials needs to ensure that equipment subsystems can meet the requirements for contamination control, perhaps similar to or even more stringent than those applied in the medical device industry. Furthermore, clean manufacturing environments will need to be designed from the ground up for scale-up manufacturing of 3D-printed biomaterials. New sterilization, packaging, and shipping procedures for 3D-printed biomaterials need to be developed to ensure product integrity at the point of use.

3.6.3 Concerns Related to Regulatory Issues

Regulatory issues are specific to a particular country and should be checked for that particular region first. In the United States, medical devices are classified under three categories based on the risk to the patient: Class I, Class II, and Class III. Class I devices are typically the lowest risk (e.g., a tongue depressor). If there is a predicate device, then many devices are classified under Class II, where the new device is substantially equivalent to the previous device, which has already been cleared by the U.S. Food and Drug Administration. Regulatory paths for the Class III or high-risk devices are quite challenging as well as time-consuming. In most cases, regulation follows technology development, and therefore, many times there is a lag in appropriate regulations for certain types of devices.

This is the case for 3D printing and its applications to human health. Although the technology for patient-specific implants is available, no clear regulatory path is established toward immediate utilization of this technology. This is also because the risk associated with the use of this technology is yet to be fully understood and evaluated. A TE graft with tailored porosity and chemistry can be fabricated in hours; however, the scientific factors that need to be verified to confirm that this will perform the same way as before are still evolving. Moreover, the advent of different 3D printing technologies is also making this evaluation more complex. Finally, the needs of specific patients are different. A clear understanding of the biomechanical and biological issues for different parts will eventually help develop regulatory pathways for diverse clinical applications of 3D-printed parts. A logical approach will establish regulatory paths for low-risk devices, such as specialty surgical tools, and then slowly move toward the high-risk devices such as patient-matched implants. It is envisioned that such trends will follow in the coming years in the United States as well as other parts of the world.

3.6.4 Material-Specific Machines

Unlike CM processes such as forging or investment casting, many 3D printing processes are different in terms of the materials that can be used and the delivery of materials. For example, a polymer-based 3D printing process will not work for metals or ceramics and vice versa. This makes process optimization very important for different materials that will be used with a specific printer. The user must know what material will be used and what kind of resolution is needed before selecting a 3D printer. This limitation is the main reason why different companies have multiple 3D printers for a variety of applications.

3.6.5 Future Trends

3D printing of multi-material structures is one of the most promising future trends. Different materials at different locations in a structure can offer properties that cannot be achieved using monolithic structures. These materials can only be bonded to each other or alloyed during the part building, allowing for compositional variations in the part. For 3D-printed biomaterials, a simple example can help us understand the power of multi-material 3D printing. For total hip prosthesis, implants are made with either Ti_6Al_4V or CoCrMo alloys. To enhance bone–tissue integration, implants are sometimes coated with porous Ti or Ta metal or calcium phosphate-based ceramics. However, different processing strategies are utilized for the addition of coatings. Using multi-material 3D printing, a calcium phosphate or metal coated hip stem can be directly manufactured from the same machine that can also produce an uncoated sample in the very next run. Therefore, in addition to flexibility in design modifications, the potential of incorporating compositional variations will allow future innovations in advanced materials and structures using 3D printing.

For bioprinting, multi-material structures include cells that can be printed along with organic or inorganic materials to make a fully functional scaffold for TE [39]. Such an approach is gaining considerable attention from a number of researchers, though there are many challenges that need careful attention. For example, cell viability will be key for successful bioprinting operation. Shelf-life of materials with and without cells is another significant issue. Also, due to significant variation in the stiffness of the cells and materials to be deposited, careful process optimization is necessary to build scaffolds that can be used for further in vitro or in vivo analysis.

Another important criterion is vascularization or new blood vessel network formation, commonly known as angiogenesis. Figure 3.5 shows new blood vessel formation in 3D-printed-doped tricalcium phosphate TE scaffolds in a rat distal femur model. 3D-printed TE scaffolds should support angiogenesis for functional tissue formation to enhance healing in vivo [40–46].

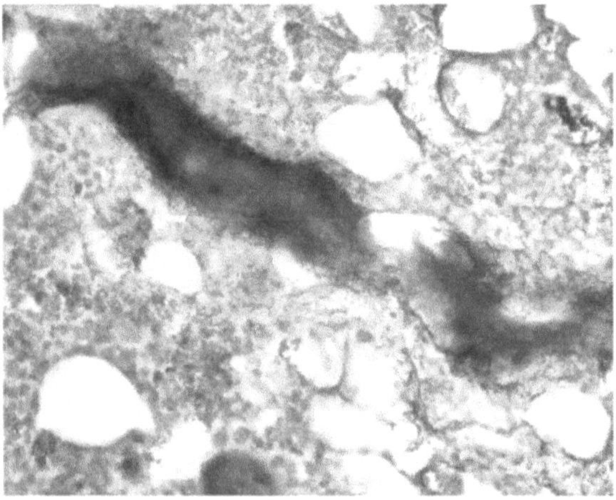

Figure 3.5 New blood vessel formation in 3D-printed-doped tricalcium phosphate TE scaffolds implanted in a rat distal femur model.

Finally, 3D printing combined with drug or protein delivery is becoming popular and is expected to gain more attention in the future. Drug-/protein-loaded biomaterials with a controlled delivery option can be used to treat many diseases [45]. In particular, specific drug combinations that are patient-matched and unique can be loaded into a single 3D-printed drug delivery device. However, drug–material and drug–drug interactions during 3D printing are complex topics and will require significant process optimization before actual applications can be realized and approved for patient use.

3.7 Conclusion

The processing of biopolymers and polymer composites through three-dimensional (3D) printing has been gaining a lot of attention, because of its multiple applications in a wide variety of industries. In fact, the exploration of the areas of applicability of this technique remains an active area of research. With the latest 3D printers capable of printing materials with a widely contrasting mechanical behavior and with complex geometries at micrometer resolutions, the potential of this technology has grown. This technology now offers the possibility of creating complex topologies with fine features, composed of a multitude of materials with variable mechanical properties on a large scale, thus allowing the manufacture of devices with an unprecedented multifunctional performance [47, 48]. Particularly in medical applications, in the last decade there has been a growing interest in the development of medical systems with bioactive properties, which have been possible thanks to the technological development of 3D printing as well as biopolymers, drugs, natural fillers and implementation in vitro or in vivo of cells in the formulated materials, which have generated a series of implants with outstanding mechanical properties and exceptional biocompatibility, adjusted to the exact dimensions of each implant for each patient. Likewise, because of the greater surface area of these devices, the cells are able to regenerate the tissues at a faster rate, thus allowing the patients to recover quickly, and avoiding a subsequent operation to remove the implant, since these implants are degraded in the time, while this one fulfills the mission for which it was developed. Finally, this type of implant also promises to be a controlled-drug-release device, while the polymer material used is degraded.

Acknowledgements

The authors would like to thank Consejo Nacional de Investigaciones Científicas y Técnicas (CONICET) (Postdoctoral fellowship internal PDTS-Resolution 2417) and Universidad Nacional de Mar del Plata (UNMdP) for the financial support, and to Dr. Mirian Carmona-Rodríguez.

References

1. Cheung, H. Y., Lau, K. T., Lu, T. P., and Hui, D. (2007). A critical review on polymer-based bio-engineered materials for scaffold development, *Compos. Part B Eng.*, **38**(3), pp. 291–300.

2. Khalil, S., Nam, J., and Sun, W. (2005). Multi-nozzle deposition for construction of 3D biopolymer tissue scaffolds, *Rapid Prototyping J.*, **11**(1), pp. 9–17.
3. Hofmann, M. (2014). 3D printing gets a boost and opportunities with polymer materials, *ACS Macro Lett.*, **3**(4), pp. 382–386.
4. Hull, C. W. (1986). Apparatus for production of three-dimensional objects by stereolithography, US patent 4,575,330.
5. 3D Systems Inc, http://www.3dsystems.com/, accessed on 03/27/2017.
6. Made in Space, http://www.madeinspace.us/#section-portal, accessed on 03/27/2017.
7. Manufacturing Demonstration Facility, Oak Ridge National Laboratory, http://web.ornl.gov/sci/manufacturing/media/news/3d-car/, accessed on 03/27/2017.
8. Deckard, C. R. (1989). Method and apparatus for producing parts by selective sintering, US patent 4,863,538.
9. Sachs, E. M., Haggerty, J. S., Cima, M. J., and Williams, P. A. (1993). Three-dimensional printing techniques, US patent 5,204,055.
10. Crump, S. S. (1992). Apparatus and method for creating three-dimensional objects, US patent 5,121,329.
11. Stratasys Ltd., http://www.stratasys.com, accessed on 03/27/2017.
12. Sanders, R. C., Forsyth, J. L., and Philbrook, K. F. (1998). 3-D model making, US patent 5,740,051.
13. Sobral, J. M., Caridade, S. G., Sousa, R. A., Mano, J. F., and Reis, R. L. (2011). Three-dimensional plotted scaffolds with controlled pore size gradients: Effect of scaffold geometry on mechanical performance and cell seeding efficiency, *Acta Biomater.*, **7**(3), pp. 1009–1018.
14. Wu, C., Luo, Y., Cuniberti, G., Xiao, Y., and Gelinsky, M. (2011). Three-dimensional printing of hierarchical and tough mesoporous bioactive glass scaffolds with a controllable pore architecture, excellent mechanical strength and mineralization ability, *Acta Biomater.*, **7**(6), pp. 2644–2650.
15. Seyednejad, H., Gawlitta, D., Kuiper, R. V., de Bruin, A., van Nostrum, C. F., Vermonden, T., Dhert, W. J. A., and Hennink, W. E. (2012). In vivo biocompatibility and biodegradation of 3D-printed porous scaffolds based on a hydroxyl-functionalized poly(ε-caprolactone), *Biomaterials*, **33**(17), pp. 4309–4318.
16. Fu, Q., Saiz, E., and Tomsia, A. P. (2011). Direct ink writing of highly porous and strong glass scaffolds for load-bearing bone defects repair and regeneration, *Acta Biomater.*, **7**(10), pp. 3547–3554.

17. Doraiswamy, A., Narayan, R. J., Harris, M. L., Qadri, S. B., Modi, R., and Chrisey, D. B. (2007). Laser microfabrication of hydroxyapatite-osteoblast-like cell composites, *J. Biomed. Mater. Res. A*, **80**(3), pp. 635–643.

18. Guillotin, B., Souquet, A., Catros, S., Duocastella, M., Pippenger, B., Bellance, S., Bareille, R., Rémy, M., Bordenave, L., Amédée, J., and Guillemot, F. (2010). Laser assisted bioprinting of engineered tissue with high cell density and microscale organization, *Biomaterials*, **31**(28), pp. 7250–7256.

19. Williams, J. M., Adewunmi, A., Schek, R. M., Flanagan, C. L., Krebsbach, P. H., Feinberg, S. E., Hollister, S. J., and Das, S. (2005). Bone tissue engineering using polycaprolactone scaffolds fabricated via selective laser sintering, *Biomaterials*, **26**(23), pp. 4817–4827.

20. Duan, B., Wang, M., Zhou, W. Y., Cheung, W. L., Li, Z. Y., and Lu, W. W. (2010). Three-dimensional nanocomposite scaffolds fabricated via selective laser sintering for bone tissue engineering, *Acta Biomater.*, **6**(12), pp. 4495–4505.

21. Lan, P. X., Lee, J. W., Seol, Y. J., and Cho, D. W. (2009). Development of 3D PPF/DEF scaffolds using micro-stereolithography and surface modification, *J. Mater. Sci. Mater. M.*, **20**(1), pp. 271–279.

22. Ronca, A., Ambrosio, L., and Grijpma, D. W. (2013). Preparation of designed poly (D, L-lactide)/nanosized hydroxyapatite composite structures by stereolithography, *Acta Biomater.*, **9**(4), pp. 5989–5996.

23. Kalita, S. J., Bose, S., Hosick, H. L., and Bandyopadhyay, A. (2003). Development of controlled porosity polymer-ceramic composite scaffolds via fused deposition modeling, *Mat. Sci. Eng. C*, **23**(5), pp. 611–620.

24. Bose, S., Darsell, J., Hosick, H. L., Yang, L., Sarkar, D. K., and Bandyopadhyay, A. (2002). Processing and characterization of porous alumina scaffolds, *J. Mater. Sci. Mater. M.*, **13**(1), pp. 23–28.

25. Darsell, J., Bose, S., Hosick, H. L., and Bandyopadhyay, A. (2003). From CT scan to ceramic bone graft, *J. Am. Ceram. Soc.*, **86**(7), pp. 1076–1080.

26. Russias, J., Saiz, E., Deville, S., Gryn, K., Liu, G., Nalla, R. K., and Tomsia, A. P. (2007). Fabrication and in vitro characterization of three-dimensional organic/inorganic scaffolds by robocasting, *J. Biomed. Mater. Res. A*, **83**(2), pp. 434–445.

27. ASTM International (2003), Standard Test Methods for Conductivity Type of Extrinsic Semiconducting Materials, available online at http://www. astm.org/Standards/F42.htm.

28. Colosi, C., Costantini, M., Latini, R., Ciccarelli, S., Stampella, A., Barbetta, A., Massimi, M., Conti, L., and Dentini, M. (2014). Rapid prototyping of chitosan-coated alginate scaffolds through the use of a 3D fiber deposition technique, *J. Mater. Chem. B*, **2**(39), pp. 6779–6791.
29. Hoque, M. E., Chuan, Y. L., and Pashby, I. (2012). Extrusion based rapid prototyping technique: An advanced platform for tissue engineering scaffold fabrication, *Biopolymers*, **97**(2), pp. 83–93.
30. Honda, M., Yada, T., Ueda, M., and Kimata, K. (2000). Cartilage formation by cultured chondrocytes in a new scaffold made of poly (L-lactide-ε-caprolactone) sponge, *J. Oral Maxil. Surg.*, **58**(7), pp. 767–775.
31. Inzana, J. A., Olvera, D., Fuller, S. M., Kelly, J. P., Graeve, O. A., Schwarz, E. M., Kates, S. L., and Awad, H. A. (2014). 3D printing of composite calcium phosphate and collagen scaffolds for bone regeneration, *Biomaterials*, **35**(13), pp. 4026–4034.
32. Balazic, M., Kopac, J., Jackson, M. J., and Ahmed, W. (2007). Review: Titanium and titanium alloy applications in medicine, *Int. J. Nano Biomater.*, **1**(1), pp. 3–34.
33. Woodruff, M. A., Lange, C., Reichert, J., Berner, A., Chen, F., Fratzl, P., Schantz, J.-T., and Hutmacher, D. W. (2012), Bone tissue engineering: From bench to bedside, *Mater. Today*, **15**(10), pp. 430–435.
34. Reichert, J. C., Cipitria, A., Epari, D. R., Saifzadeh, S., Krishnakanth, P., Berner, A., Woodruff, M. A., Schell, H., Mehta, M., Schuetz, M. A., Duda, G. N., and Hutmacher, D. W. (2012). A tissue engineering solution for segmental defect regeneration in load-bearing long bones, *Sci. Transl. Med.*, **4**(141), p. 141ra93.
35. Melchels, F. P., Domingos, M. A., Klein, T. J., Malda, J., Bartolo, P. J., and Hutmacher, D. W. (2012). Additive manufacturing of tissues and organs, *Prog. Polym. Sci.*, **37**(8), pp. 1079–1104.
36. Hutmacher, D. W. (2013). A road map for a tissue engineering concept for restoring structure and function after limb loss, *J. Biomed. Mater. Res. A*, **24**(11), pp. 2659–2663.
37. Eshraghi, S., and Das, S. (2012). Micromechanical finite-element modeling and experimental characterization of the compressive mechanical properties of polycaprolactone–hydroxyapatite composite scaffolds prepared by selective laser sintering for bone tissue engineering, *Acta Biomater.*, **8**(8), pp. 3138–3143.
38. Eshraghi, S., and Das, S. (2010). Mechanical and microstructural properties of polycaprolactone scaffolds with one-dimensional, two-

dimensional, and three-dimensional orthogonally oriented porous architectures produced by selective laser sintering, *Acta Biomater.*, **6**(7), pp. 2467–2476.

39. Ringeisen, B. R., Pirlo, R. K., Wu, P. K., Boland, T., Huang, Y., Sun, W., Hamid, Q., and Chrisey, D. B. (2013). Cell and organ printing turns 15: Diverse research to commercial transitions, *MRS Bull.*, **38**(10), pp. 834–843.

40. Fielding, G., and Bose, S. (2013). SiO_2 and ZnO dopants in three-dimensionally printed tricalcium phosphate bone tissue engineering scaffolds enhance osteogenesis and angiogenesis in vivo, *Acta Biomater.*, **9**(11), pp. 9137–9148.

41. Tarafder, S., Dernell, W. S., Bandyopadhyay, A., and Bose, S. (2015). SrO-and MgO-doped microwave sintered 3D printed tricalcium phosphate scaffolds: Mechanical properties and in vivo osteogenesis in a rabbit model, *J. Biomed. Mater. Res. B*, **103**(3), pp. 679–690.

42. Tarafder, S., Davies, N. M., Bandyopadhyay, A., and Bose, S. (2013). 3D printed tricalcium phosphate bone tissue engineering scaffolds: Effect of SrO and MgO doping on in vivo osteogenesis in a rat distal femoral defect model, *Biomater. Sci.*, **1**(12), pp. 1250–1259.

43. Bose, S., Roy, M., and Bandyopadhyay, A. (2012). Recent advances in bone tissue engineering scaffolds, *Trends Biotechnol.*, **30**(10), pp. 546–554.

44. Bose, S., Fielding, G., Tarafder, S., and Bandyopadhyay, A. (2013). Understanding of dopant-induced osteogenesis and angiogenesis in calcium phosphate ceramics, *Trends Biotechnol.*, **31**(10), pp. 594–605.

45. Bose, S., and Tarafder, S. (2012). Calcium phosphate ceramic systems in growth factor and drug delivery for bone tissue engineering: A review, *Acta Biomater.*, **8**(4), pp. 1401–1421.

46. Tarafder, S., and Bose, S. (2014). Polycaprolactone-coated 3D printed tricalcium phosphate scaffolds for bone tissue engineering: In vitro alendronate release behavior and local delivery effect on in vivo osteogenesis, *ACS Appl. Mater. Interfaces*, **6**(13), pp. 9955–9965.

47. Dimas, L. S., Bratzel, G. H., Eylon, I., and Buehler, M. J. (2013). Tough composites inspired by mineralized natural materials: Computation, 3D printing, and testing, *Adv. Funct. Mater.*, **23**(36), pp. 4629–4638.

48. Ge, Q., Qi, H. J., and Dunn, M. L. (2013). Active materials by four-dimension printing, *Appl. Phys. Lett.*, **103**(13), p. 131901.

Chapter 4

Proteins and Their Novel Applications

Tanvir Arfin, Shoeb Athar, and Stephy Rangari

Environmental Materials Division,
CSIR-NEERI, Nehru Marg, Nagpur, Maharashtra, India

tanvirarfin@ymail.com

Proteins are large macromolecules that have long chains of amino acids residues known as polypeptides. A broad range of functions performed by proteins includes catalysis of metabolic reactions, response to stimuli, DNA replication, and transport of molecules from one place to another. In the past several years, researchers have successfully characterized the broad variety of proteins with different structural design catering the functional needs of living beings at the molecular scale. This has led to the active development of numerous innovative protein applications in various forms. This chapter discusses different aspects of proteins along with their novel applications. Starting with the historical background, structure of protein, their types, synthesis, and cellular functions are briefly discussed. Nutritional importance of protein is also emphasized. The latter section of the chapter deals in detail with novel protein applications in the

Handbook of Biopolymers: Advances and Multifaceted Applications
Edited by Shakeel Ahmed, Suvardhan Kanchi, and Gopalakrishnan Kumar

ISBN 978-981-4800-17-4 (Hardcover), 978-0-429-02475-7 (eBook)
www.panstanford.com

form of carrier proteins, green fluorescent protein (GFP), custom-tailored fusion proteins, functional ingredients, protein biochips, and protein nanoparticles. Significant outcomes and results are also highlighted in this regard.

4.1 Introduction

Proteins are large macromolecules containing a large number of chains of amino acids residues, called polypeptides. They carry out large functions inside the living organisms, along with DNA replication and transportation of particles from different places. Proteins vary with each other mainly on their arrangement of amino acids, which is encoded by the genes of their nucleotide arrangement which is a basic unit of cell, and which mostly results in protein folding into a particular three-dimensional format that is useful to determine its task.

A linear organic polymer is a long and straightforward chain of amino acid residues. Polypeptides containing less than 20–30 residues of amino acid are short in nature, which are rarely considered to be proteins or frequently oligopeptides as shown in Fig. 4.1. This peptide bond shows resonance property as shown in Fig. 4.2. The particular residues of amino acid are bonded with each other by a chemical bond between the amino acids and neighbors of amino acid remainder. The arrangement of residues in a protein molecule of the amino acid is explained by a gene in a nucleotide sequence, which is encoded in the molecule which carries genetic information can add selenocysteine and in some chief pyrrolysine. At the time of preparation of the protein, the residues of amino acid are improved by the posttranslational improvement which changes its various physiological characteristics and finally protein functions also.

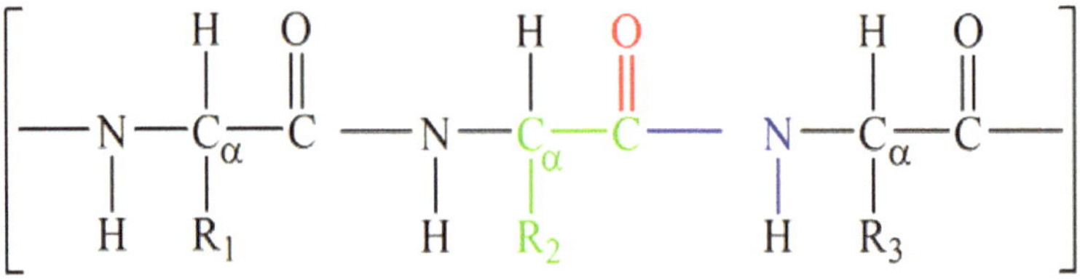

Figure 4.1 Chemical structure of the peptide bond.

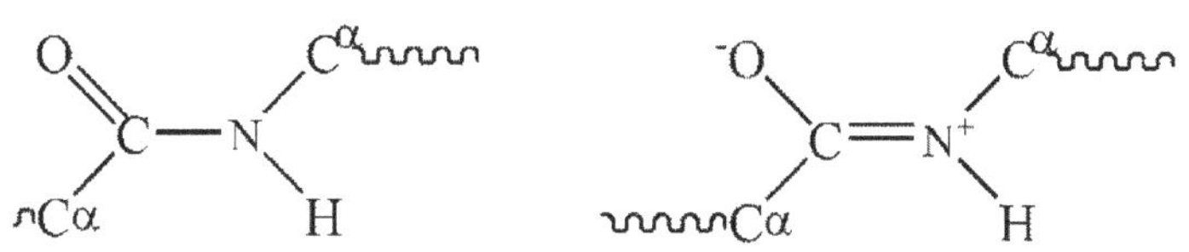

Figure 4.2 Resonance behavior in peptide bond.

Proteins can make an effort to pursue precise functions and them allying to develop constant protein complexes.

4.2 History of Protein

Proteins have long been considered a particular category of biological family molecules since 18th century. Antoine Fourcroy and others identified them by their capability to clabber or to form small clumps beneath the treatment of heat or acid [1]. An example of proteins at the time comprised egg white albumin and wheat mixtures. It was Dutch chemist Gerardus Johannes Mulder who first described proteins and the same was named by Jons Jacob Berzelius, a Swedish chemist in 1838 [2].

Carl Von Volt suggested that protein molecule was an essential factor for supporting the growth and functions of a human body since it was usually thought that "flesh builds flesh" [3].

Karl Heinrich Ritthausen expanded called protein type with recognition of glutamic acid. The understanding of proteins as a polypeptide chain of amino acid came out along with the effort of Franz Hofmeister and Hermann Emil Fischer scientist. The central part of the living organism of protein as enzymes was not accepted in 1926 after that James B. Surer suggested that every enzyme is a protein, but every protein is not an enzyme.

Limitations in purifications of proteins in large amounts posed a significant hindrance to early biochemists for study. Therefore, previous studies on proteins were countered that it could be cleaned in massive quantities such as egg white, blood vessels, several hazardous and metabolism with enzymes acquired from the abattoir. Frederick Sanger sequenced the structure of insulin in 1949. Sanger accurately found out the sequence of amino acids of insulin and therefore decisively showed that proteins are made up of the linear polymer of amino acids and the branched chains are not involved, colloids or cycols [4]. Max Perutz and Sir John Cowdery Kendrew solved protein structures of hemoglobin and myoglobin, respectively [5, 6].

4.3 Structure of Proteins

Most of the protein structures fold into a different 3D structure. The format of protein molecules pleats is called their constitutional special arrangement. With the help of chemical properties of the amino acids, proteins can fold unattended; some proteins require the extra facility of particles to pleats into their natural conditions. Biochemists usually cite four different aspects of the structure of protein.

4.3.1 Primary Structure

The structure of primary protein consists of simple and linear amino acid arrangement. A protein molecule is also termed as polyamide.

4.3.2 Secondary Structure

The hydrogen bonds can be fixed up the local structure of protein. The α-helix, β-sheet, and turns are complete modern examples of the secondary form of protein. Various structures of different regions can occur in the molecules of protein, due to the reason of secondary structure being local.

4.3.3 Tertiary Structure

The formation of a hydrophobic core and some interactions can lead to stable the tertiary structure of protein, but again with the help of salt bridges and the posttranslational modification of proteins. The word pleats are usually identical to the word "tertiary structure." The main necessary functions of the proteins are regulated by the tertiary structure.

4.3.4 Quaternary Structure

The quaternary structure is the structure formed by some different molecules of protein mainly termed as protein subunit and serve as a single protein complex.

The three broad types interact with the tertiary structure such as membrane proteins, globular proteins, and fibrous proteins in which proteins can be categorized. Most of the globular proteins are simply soluble, but some of the proteins those are globular are enzymes. A collagen is an essential part of connective tissues, or keratin, the protein ingredient for the proper growth of nails and hair. For the proper route of polarity or charged atoms to proceed across the membrane layer, membrane proteins mainly act as a receptor or support channels for them.

4.4 Types of Proteins

A single chain of polypeptide consisting of a protein molecule is called monomeric; the proteins that are composed of a large number of the polypeptide chain are known as oligomeric. By the chemical composition of proteins, they are categorized into principal classes: first class is simple proteins. These are made up of single amino acids, and the second class is conjugate proteins. These are made up of amino acids, and other presence of organic and inorganic groups, some of them are known as prosthetic groups. Glycoproteins are those which are included under conjugated proteins, and they possess carbohydrates, lipoproteins composed of lipids and nucleoproteins in which nucleic acids are included.

On the basis of classification by biological functions, proteins possess enzymes, and these enzymes are quite important to carry out the thousands of chemical reactions under the organisms; keratin, elastin, collagen, those are the essential types of structural proteins; hemoglobin and also other proteins molecule; casein a protein present in milk and some different essential nourishment required for growth; antibodies are important fragment for protection of immune system; hormones that control the whole body metabolism are protein hormones, and those proteins which can acquire mechanical work of actin and myosin are muscle of contractile proteins. The types of protein with function are described in Table 4.1.

Table 4.1 Various type of proteins

Types of protein	Function	Examples
Motility	Cells and their organelles movement	Tubulin forms microtubules to move; actin and myosin work together to move
Structural	Provides protection and strength	Keratin forms nails and hairs; collagen supports tissues
Enzymes	Enhances biochemical reactions	Catalase removes toxic hydrogen peroxide from cells
Transport	Transfers molecules from one place to another	Hemoglobin carries oxygen to body cells

4.5 Synthesis of Protein

The three different methods—genetics, physics, and biochemistry—have been devoted to a large part of the improvement of education regarding genetical information path. The main contributions of these three fields get involved in the invention of the double-helical structure of DNA by Watson and Crick in the year of 1953. Genetics have been supported the theory of coding, whereas the chemical composition of DNA was acknowledged by biochemistry. The structure of DNA proposed that how DNA could be transcribed so that the information present within it could be passed from one to other generation. The knowledge about how the information incorporated in DNA was transformed to functional proteins has become available over the disclosure of messenger RNA, transfer RNA and genetic code. All these improvements lead to the perceptive of the protein synthesis mechanism within the central dogma of molecular biology, which posits that genetic information from one generation to other is carried out from nucleic acids to protein. The first method is replication, in which parental DNA gets replicate into two daughter DNAs which have same sequences of nucleotide. The second is the transcription; in this approach, the message coded in DNA is copied accurately to the form of RNA. The RNA formed from DNA is known as messenger RNA (mRNA), and the information to the nucleotide sequence is incorporated in the parental DNA molecule. Parental DNA and the mRNA resulting from it are proportional in nature. That is why the change of code is not included in the method of transcription. Transcription may be termed as a synthesis of RNA from DNA and is DNA-dependent. The third is translation; in this step, the genetic code present on mRNA gets translated to ribosomes inside a protein consisting of a particular type of amino acids sequence. In different terms, translation is defined as the production of the primary chain of polypeptides corresponding to the recognition of mRNA. Evidently, the translation process eventually includes a code change from sequences of a nucleotide to sequences of amino acid because both of them are not harmonious. Thus, corresponding to the central dogma, the passing of information in the form of the sequence is one-directional, i.e., from DNA the information in the form of nucleotide sequence is transcribed to RNA and from RNA the information in the form of amino acids sequence is translated into protein. In other words, "DNA produce RNA, and RNA produce proteins."

4.6 Cellular Functions

Proteins are reported to be the principal performers within the cell fulfilling the obligations stated by genetically coded information [7]. Protein–protein interactions control the activity of enzymes and regulate progression via the cell cycle besides permitting the assemblage of large protein molecule complexes, which perform many intimately linked reactions with a mutual biological function. Proteins in the form of enzymes are known to catalyze some 4,000 chemical reactions, including those involved in metabolism. They are also involved in DNA manipulation in the processes of DNA repair and replication as well as transcription. Some enzymes also take part in a process called posttranslational modification, where they act upon other proteins to remove some chemical groups [8]. Excluding some specific types of RNA, a majority of biological molecules are comparatively inert elements on which the action of protiens takes place[9].

The proteome is the term for the particular expression of a set of proteins in a certain cell. The capability of proteins to firmly and explicitly bind other molecules is their primary characteristic permitting a myriad set of functions. The site of the protein in charge to bind a molecule is called "the binding site" —typically a dip on the surface. The tertiary structure of protein mediates the ability of the protein to bind other molecules defining the binding site. However, the analytic functions of the neighboring chains of amino acids also have a significant contribution. Protein binding is exceptionally robust and specific as exemplified by the binding affinity of the ribonuclease inhibitor protein to human angiogenin with a sub-femtomolar dissociation constant lesser than 10−15 M and the non-binding tendency for its amphibian homolog increase with dissociation constant (>1 M). Notably, even slightest chemical changes such as adding a unit methyl group to the binding partner may at times result in near elimination of binding [10]. Proteins bind to other proteins besides binding with small molecular substrates. On specific binding with other replicas of the same molecule, protein oligomerization takes place resulting in fibril formation, a development that frequently occurs in structural proteins consisting of globular monomers which self-assemble together resulting in the formation of rigid fibers. Structural proteins are responsible for imparting firmness and rigidity to natural elements which are otherwise fluid in nature. A majority of structural proteins such as collagen and elastin—fundamental constituents of connective tissues like cartilage—are fibrous. Another example may include keratin, which is the building block of hard or filamentous constructions, for example, hair, nails, and hooves [11]. Furthermore, nearly half of the dry weight of an Escherichia coli cell is made up of proteins, although other macromolecules such as RNA and DNA contribute merely 20% and 3% of the dry weight, respectively.

Proteins are also known for their role in cell signaling and signal transduction process. For example, extracellular proteins such as insulin take the part signal transmission from the cell of their synthesis to the cells located in remote tissues. Then there are some membrane-type proteins that function as receptors, binding signaling molecules, and stimulating a biochemical response in the cell. A large number of receptors possess an exposed binding site on the surface of the cell along with a nerve fiber domain inside the cell, likely having enzymatic action or tendency to experience a conformational transformation sensed by different proteins inside the cell [12]. The potential of attraction partner to prompt conformational alterations inside proteins facilitates the formation of extremely complicated flagging systems [13].

4.7 Novel Applications

Various types of applications are presented in the following subsections.

4.7.1 Carrier Protein

Acyl carrier proteins (ACP), renowned for their critical role in fatty acid biosynthesis, have been demonstrated to cooperate with some proteins in the fatty acids construction [14]. Apart from this, carrier proteins also play a crucial task in polyketide synthase and non-ribosomal synthase frameworks [15].

4.7.2 Biotechnology

The structural information that acyl carrier proteins—which play an instrumental role in fatty acid biosynthesis—possess paves the way for their use in several biotechnological applications. Coenzyme A (CoA) derivatives can be covalently bonded to carrier proteins where these tailored 4′-phosphopantetheine arms may perform in reactant frameworks or function to hinder the same. Therefore, carrier proteins may either be tailored in vitro for analysis or applied for engineering the biosynthesis of natural products in vitro. Besides, exclusive features are exploited in their use as a fusion tag.

Carrier proteins can be specifically tailored by a 4′-phosphopantetheinyl transferase (PPTase) with an unlimited range of its practical gatherings. Several identified PPTases must quickly take atypical CoA simple, comprising those having a covalently reinforced reported moiety [16]. The groups may move to carrier proteins where fluorescence or purification with a partiality tag might be utilized to envision them.

Incorporation of a broad range of pantetheine analogs into carrier proteins has been rendered possible by the highly permissive PPTASE-Sfp (a highly promiscuous PPTASE from surfactin biosynthesis) in vitro. This feature of PPTASE was first used to attach non-cognate amino acids onto PCPs (peptidyl carrier protein) to investigate non-ribosomal synthase (NRPS) systems domain activity [17]. Transfer of even very large CoA analogs which also include natural products with long polypeptide is the further extension of this application [18]. Modification of carrier proteins via this method permits the monitoring other proteins as well as carrier proteins in the respective pathways for their activity [19].

George et al. [20] first reported the in vivo carrier proteins labeling on cell surfaces through CoA and 4′-phosphopantetheine analogs. They expressed E. coli acyl carrier protein (ACP) as a combination with a cell surface protein—the alpha-agglutinin receptor Aga2p from Saccharomyces cerevisiae—and selectively labeled the fusion proteins on surface of cells applying Sfp and CoA-maleimide analogs. Incubation of cells communicating the ACP combination with fluorescent CoA analogs was carried out to modify the protein ranges of combination proteins fluorescently.

Recently, numerous sophisticated works have been reported where carrier protein fusions were labeled on surfaces of cells. Possibility for specific modification of carrier protein fusions has permitted for integration of multiple cell surface labels and also time-dependent cell surface labeling.

4.7.3 Engineering

Engineering of novel polyketide synthase (PKS) and NRPS is one of the first objectives in the biosynthesis of natural products. Although some researchers have worked toward this end, the function of carrier protein domains in these attempts remains between far and few. Synthesis of novel products from engineered systems has indeed been reported, but engineering, on the other hand, rigorously reduces the efficiency of biosynthesis. The activity decreases inbuilt PKS and NRPS is attributed to the disturbance in the folding and inter-domain interactions in biosynthetic pathways where a sequence of influential positions accommodated on a lone polypeptide and to the disruption in protein–protein interactions in biosynthetic pathways where every dynamic position is located on a distinct polypeptide.

4.7.4 Biosensors

The most prevalent application of protein-doped S-G materials to have evolved and is in the form of biosensors. The principal mechanism of biosensors is the careful interfacing of the bio-acknowledgment part to the physical transducer where the latter translates the natural response to a quantifiable and computable signal. Thermal, electrochemical, optical, and piezoelectric methods are the usual transduction schemes employed in biosensors. A greater part of S-G construct biosensors is situated in light of electrochemical and optical transduction plot. Systems, including surface plasmon reverberation and quartz precious stone, are not employed extensively, probably, due to the suitability of these methods for analytes with higher mass, which is likely incompatible with detection using S-G trapped proteins because of remoteness of caught proteins to the bigger analytes [21].

4.7.5 Electrochemical Biosensors

S-G derivative materials have garnered researchers' attention by offering a resourceful approach for the preparation of modified electrodes [22]. The key advantage that S-G capability bestows upon electrochemical detecting is the plausibility of a class of electrode configurations. For instance, the possibility of preparation of electrodes containing directing species, redox average people, some of the chemicals, where each one may be employed to regulate the electrode features finely, is always there. Regular leading species are graphite and palladium; however, as of late the concentration has moved toward the joining of gold colloids into S-G glasses [23]. Modulation of enzyme activity is usually achieved by osmosis and polymers. For example, Heller and colleagues incorporated polycationic polymers into S–G materials and observed an improvement in the performance of flavoproteins [24]. Notwithstanding this, Wang et al. [25] have demonstrated that fusing copolymers into silica-based glasses can altogether improve the captured glucose oxidase movement for recognizing glucose amperometrically.

4.7.6 Optical Biosensors

Working of optical biosensors depends on the ensnaring biomolecules containing chromophoric or fluorescent gatherings, or co-entrapping a biomolecule with a chromophore or fluorophore receptive to the protein-encouraged response. Rather, bioluminescence might be misused as a signal for reporting the present species influencing the behavior of luminescent proteins like aequorin [26]. Optical sensors identify analytes given their assimilation or outflow of light. However, reconciliation of fluorescence extinguishing plans into the sensing system opens up the likelihood to widen the extent of analytes that may incorporate the non-engrossing classification those conservatives the disintegrate qualities of the caught fluorescent species.

4.7.7 Bioaffinity Chromatography

Change of stationary stages for chromatography is a rising utilization of S-G determined materials. Both polar and non-polar S-G materials are connected to get ready standard and turned around stable steps. This biocomposite are likewise used to get ready resistant fondness segments [27]. In the blink of an eye, particulate segments alongside substantial sections and open tubular segments are the three distinct arrangements of sol–gel-based segment reported in a few studies. Particulate segments are set up for the next expansion of S-G powders into the segment with the help of a dissolvable slurry strategy [28]. The got sections, as a rule, have generous void volumes due to voids between particles and also pores inside the particles. This guarantees unrivaled stream rates as a result of interparticle division with higher surface range because of pores of nanoscale measurements.

4.7.8 Solid-Phase Extraction

Test pre-treatment through active stage extraction/miniaturized scale extraction (SPME) is a recognized utilization of S-G determined materials. Pawliszyn presented SPME coatings as a procedure for pre-convergence of picked analytes [29]. The technique includes setting a film-covered fiber into a specimen and assurance of apportioning of analytes by the particular parcel co-effective for the analyte. The minute balance is established between the film and the analyte sample, the film is confined, and dissolvable extraction or vanishing extricates the analyte from the film for further investigation utilizing conventional strategies like GC/MS or LC/MS. Of late, a majority of SPME coatings are obtained from materials such as polydimethylsiloxane (PDMS) or carbowax providing for selectively extracting the non-polar compounds from samples apart from other coatings like polyacrylate that are used for polar and ionic compounds.

4.7.9 Protein Biochips

A biochip can be alluded to as an assembly of scaled-down spots additionally called microarray, which is sorted out on a high substrate permitting the execution of different tests at the same time to accomplish high-throughput screenings [30]. A protein biochip may be defined as the protein microarray having a well-ordered arrangement of various proteins for example antibodies on the chip with known coordinates. Such proteins function as capture agents for chosen proteins. To render the captured proteins detectable, they are directly labeled with a fluorophore or as in the case of antibody chips, a layer of labeled antibodies is added to make a sandwich. The binding location of a protein discloses the type of antibody it binds with and thereby revealing its functional identity. Moreover, the signal strength from that position indicates the concentration of protein. With the advantage of high throughput, antibody microarrays can theoretically examine thousands of proteins in a sample droplet, serum for example. Protein biochip systems have the ability to automate protein identification, protein quantification, and biomolecular interactions studies. Protein biochips that are fundamentally not microarrays are also being employed for biological and medical research, which makes the investigation of biomolecular interactions or selection of protein from protein mixtures simpler and easier. Integrated protein biochip systems consist of chips and reagents, a chip processing instrument, a scheme for captured proteins detection and quantification and software capable of analyzing large amounts of data [31].

4.7.10 Protein Nanoparticles

Protein nanoparticles have been reported for delivery applications of almost all kind of therapeutics, including small-molecule drugs, peptides, nucleic acids, and proteins. Among these, therapeutic protein and peptides are most prone to functional loss due to degradation or other mechanism and protein NPs can be most beneficial for their delivery. Protein or peptides to be delivered can either be loaded onto these particles or directly be used as a building material for the fabrication of these particles [32].

4.8 Conclusions

The chapter discussed the historical background, structure, and cellular functions of proteins precisely and their novel protein applications in details. Proteins are long macromolecules involving at least one protracted chains of amino corrosive deposits called peptide. Proteins can be considered as molecular devices with nanometric dimensions whereupon biological function is exerted. Being the building block of all cells in all living organisms of every kingdom, they execute the effective course of life maintenance, replication, defense, and reproduction. Four distinct parts of protein structure are sorted as an essential, auxiliary, tertiary, and quaternary structure by natural chemists. Proteins as enzymes catalyze chemical reactions in the cell. Besides, they are instrumental in the cell signaling as well as signal transduction process and signal transduction. Myriad novel protein applications have gradually evolved over decades by virtue of carrier proteins, proteins embodied inside sol–gel determined materials, GFP, custom-tailored fusion proteins, functional ingredients, protein biochips, protein nanoparticles, etc. These applications include bio-sensing, bio-regulation, therapeutics, bio-catalysis, pH detection, reporter genes, fusion tags, fluorescence resonance energy transfer, photo-bleaching, biophysics, solid phase extraction, and bio-affinity chromatography. However, with current research impetus on proteins and their novel applications, new apertures are likely to be revealed rendering their further exploration a simpler task with tunable and foreseeable outcomes.

References

1. Osbone, T. B. (1909). *The Vegetable Proteins*, 2nd ed. (Longmans Green and Co., London).
2. Harold, H. (1951). Origin of the word 'protein', *Nature*, **168**, pp. 244–244.
3. Bischoff, T. L. W., and Volt, C. (1860). *Die gesetze der ernahrung des fleischfressers durch neue untersuchungen festgestellt*, (Winter, Leipzig Heidelberg).
4. Sanger, F. (1949). The terminal peptides of insulin, *Biochem. J.*, **45**, pp. 563–574.
5. Muirhead, H., and Perutz, M. (1963). Structure of haemoglobin. A three–dimensional Fourier synthesis of reduced human haemoglobin at 5.5 a resolution, *Nature*, **199**, pp. 633–638.
6. Kendrew, J., Bolo, G., Dintzis, H., Parrish, R., Wyckoff, H., and Phillips, D. (1958). A three-dimensional model of the myoglobulin molecule obtained by X-ray analysis, *Nature*, **181**, pp. 662–666.
7. Lodish, H., Berk, A., Matsudaira, P., Kaiser, C. A., Krieger, M., Scott, M. P., Zipurksy, S. L., and Darnell, J. (2004). *Molecular Cell Biology*, 5th ed. (WH Freeman and Company, New York).
8. Bairoch, A. (2000). The ENZYME database in 2000, *Nucleic Acids Res.*, **28**, pp. 304–305.
9. Voet, D., and Voet, J. G. (2004). *Biochemistry*, vol. 1, 3rd ed. (Wiley: Hoboken, New Jersey).
10. Sankaranarayanan, R., and Moras, D. (2001). The fidelity of the translation of the genetic code, *Acta Biochim. Pol.*, **48**, pp. 323–335.
11. Christopher, K. M., and Van Hold, K. E. (1996). *Biochemistry*, 2nd ed. (Benjamin/Cummings, Menlo Park, California).
12. Branden, C. I., and Tooze, J. (1999). *Introduction to Protein Structure*, 2nd ed. (Garland Publishing, New York).
13. Brosnan, J. T. (2003). Interorgan amino acid transport and its regulation, *J. Nutr. Nutrition*, **133**, pp. 2068S–2072S.

14. Rawlings, M., and Cronan, J. E. (1992). The gene encoding *Escherichia coli* acyl carrier protein lies within a cluster of fatty acid biosynthetic genes, *J. Biol. Chem.*, **267**, pp. 5751–5754.
15. Hill, A. M. (2006). The biosynthesis, molecular genetics and enzymology of the polyketide-derived metabolites, *Nat. Prod. Rep.*, **23**, pp. 256–320.
16. La Clair, J. J., Foley, T. L., Schegg, T. R., Regan, C. M., and Burkart, M. D. (2004). Manipulation of carrier proteins in antibiotic biosynthesis, *Chem. Biol.*, **11**, pp. 195–201.
17. Belshaw, P. J., Walsh, C. T., and Stachelhaus, T. (1999). Aminoacyl-CoAs as probes of condensation domain selectivity in nonribosomal peptide synthesis, *Science*, **284**, pp. 486–489.
18. Sieber, S. A., Walsh, C. T., and Marahiel, M. A. (2003). Loading peptidyl-coenzyme A onto peptidyl carrier proteins: A novel approach in characterizing macrocyclization by thioesterase domains, *J. Am. Chem. Soc.*, **125**, pp. 10862–10866.
19. Yin, J., Straight, P. D., McLoughlin, S. M., Zhou, Z., Lin, A. J., Golan, D. E., and Walsh, C. T. (2005). Genetically encoded short peptide tag for versatile protein labeling by Sfp phosphopantetheinyl transferase, *Proc. Natl. Acad. Sci. U. S. A.*, **102**, pp. 15815–15820.
20. George, N., Pick, H., Vogel, H., Johnsson, N., and Johnsson, K. (2004). Specific labeling of cell surface proteins with chemically diverse compounds, *J. Am. Chem. Soc.*, **126**, pp. 8896–8897.
21. Jin, W., and Brennan, J. D. (2002). Properties and applications of proteins encapsulated within sol–gel derived materials, *Anal. Chim. Acta*, **461**, pp. 1–36.
22. Walcarius, A. (2001). Electrochemical applications of silica-based organic-inorganic hybrid materials, *Chem. Mater.*, **13**, pp. 3351–3372.
23. Niu, J., and Lee, J. Y. (2000). Bulk-modified amperometric biosensors for hypoxanthine based on sol–gel technique, *Sens. Actuator B Chem.*, **62**, pp. 190–198.
24. Heller, J., and Heller, A. (1998). Loss of activity or gain in stability of oxidases upon their immobilization in hydrated silica: Significance of the electrostatic interactions of surface arginine residues at the entrances of the reaction channels, *J. Am. Chem. Soc.*, **120**, pp. 4586–4590.
25. Wang, B., Li, B., Deng, Q., and Dong, S. (1998). Amperometric glucose biosensor based on sol-gel organic-inorganic hybrid material, *Anal. Chem.*, **70**, pp. 3170–3174.

26. Blyth, D. J., Poynter, S. J., and Russell, D. A. (1996). Calcium biosensing with a sol–gel immobilized photoprotein, *Analyst*, **121**, pp. 1975–1978.

27. Tanaka, N., Kobayashi, H., Nakanishi, K., Minakuchi, H., and Ishizuka, N. (2001). Peer reviewed: Monolithic LC columns, *Anal. Chem.*, **73**, pp. 420A–429A.

28. Bronshtein, A., Aharonson, N., Turniansky, A., and Altstein, M. (2000). Sol-gel-based immunoaffinity chromatography: Application to nitroaromatic compounds, *Chem. Mater.*, **12**, pp. 2050–2058.

29. Arthur, C. L., and Pawliszyn, J. (1990). Solid phase microextraction with thermal desorption using fused silica optical fibers, *Anal. Chem.*, **62**, pp. 2145–2148.

30. Rusmini, F., Zhong, Z., and Feijen, J. (2007). Protein immobilization strategies for protein biochips, *Biomacromolecules*, **8**, pp. 1775–1789.

31. Sage, R. F. (2004). The evolution of C4 photosynthesis, *New Phytol.*, **161**, pp. 341–370.

32. Estrada, L. H., and Champion, J. A. (2015). Protein nanoparticles for therapeutic protein delivery, *Biomater. Sci.*, **3**, pp. 787–799.

Chapter 5

Chitin and Chitosan: The Defense Booster in Agricultural Field

Vijayalakshmi Kumar, K. Sangeetha, P. Ajitha, S. Aisverya, S. Sashikala, and P. N. Sudha

PG & Research Department of Chemistry,
D.K.M. College for Women, Vellore, Tamil Nadu, India

drparsu8@gmail.com

5.1 Introduction

Chitosan biopolymer has been proven to induce abiotic and biotic stress tolerance in various horticultural commodities, and it is also utilized to protect edible products and for plant growth stimulation [1]. Since 1990, the studies on chitin and chitosan have been intensified due to their excellent properties, such as biodegradation in the human body, controlling plant diseases, immunological, antibacterial, antifungal, wound-healing activity. Hence, the chitin and chitosan are receiving greater attention as novel functional materials. Despite their interesting

Handbook of Biopolymers: Advances and Multifaceted Applications
Edited by Shakeel Ahmed, Suvardhan Kanchi, and Gopalakrishnan Kumar

ISBN 978-981-4800-17-4 (Hardcover), 978-0-429-02475-7 (eBook)
www.panstanford.com

biological properties, in recent studies, especially the chitosan biopolymer has been found to be a good candidate, as a support material for the induction of defensive system of plants and plant-growth promotion, which plays key roles in agricultural applications. The eliciting activities of chitin and chitosan fragments lead to a variety of defense responses in host plants in response to microbial infections, including pathogen-related proteins and protienase inhibitors, accumulation of phytoalexins, lignin synthesis, and cellulose formation. Primarily, the agricultural uses of chitosan for plant defense and yield increase were based on how this glucosamine polymer influences the biochemistry and molecular biology of the plant cell. During the 20th century, especially in agricultural field all over the world, chitosan has been used in agriculture in order to feed the ever-increasing population by increasing the productivity per unit area of land used for crop production [2]. Chitosan nanoparticles, the recent innovatively modified form of chitosan, have been utilized as a valuable delivery system for pesticides, fertilizers, herbicides, and micronutrients for crop growth promotion by a balanced and sustained nutrition.

5.1.1 Biopolymers

Polymers also referred to as intelligent environmental-friendly polymers are produced naturally from living organisms [3], plants, and animals and are often biodegradable and non-toxic. Biopolymers and their derivatives are abundant, diverse, and very important for life. Since the biopolymers made from plant materials can be grown indefinitely, these can be sustainable, carbon neutral, and renewable, and some are also compostable. Almost all the biopolymeric material accounts for the greater part of the human body and as well as the ecosphere. In most organisms, these biopolymers contribute the major fraction of cellular dry matter, and the living matter is able to synthesize a wide range of different polymers [4]. The slight difference between the polymers and the biopolymers is that the polymers have random and simpler structures, whereas the biopolymers have a well-defined primary structure, which folds into compact shapes.

Biopolymers are strings or sequences of monomeric units or monomers like every polymer, and in many cases these strings are linear, but sometimes they are closed and circular, branched, or even cross-linked. Biopolymers have emergent properties associated with their hierarchical structures [5]. Based on their structure and the used monomeric units, biopolymers have been mainly classified as follows: (i) long-chain polynucleotides (RNA and DNA) composed of 13 or more nucleotide monomers, (ii) polysaccharides often containing linear bonded polymeric carbohydrate structures, and (iii) polypeptides composed of short polymers of amino acids [6–9]. Biopolymers may be produced naturally or by chemical synthesis [10].

The natural biopolymers produced within living organisms as structural components of tissues include mainly proteins and polysaccharides. The commonly used polysaccharides include cellulose, hyaluronate, chitin, chitosan, alginic acid with its salts, and heparin [11]. Polymers produced from both living organisms and chemical synthesis are widely used in medical applications, but their particular utilities are determined by specific properties and possibilities of shaping the final products [12]. Synthetic biopolymers' classification emphasizes that plastics are more or less susceptible to active biological environment. Non-degradable synthetics are willingly used in various types of surgeries (i.e., plastic, reconstructive, vascular, or trauma), and the most popular are silicones, polyethylene (PE), polypropylene (PP), polyamide (PA), polyurethane (PU), polymethacrylates (PMA), polytetrafluoroethylene (PTFE), polycarbonates (PC), acrylic resins, and polyvinyl chloride (PVC).

Certain examples of biopolymers are chitosan, chitin, cellulose, starch, DNA, RNA, proteins, and peptides in which the monomeric units are the sugars, amino acids, and nucleotides. Depending on the synthesis process, biodegradable polymers can also be classified in various ways, which have already been proposed in the literature [13]. According to their chemical structure, biopolymers can be mainly classified into eight major classes: (i) polymers such as starch, cellulose, glucans, mannans, chitin, and chitosan [14], (ii) polyamides such as proteins and poly(α- or β-) amino acids, (iii) polynucleotides such as nucleic acids, (iv) organic polyoxoesters such as poly hydroxy alkanoic acids, polymalic acid, and cutin, (v) inorganic polyesters such as polyphosphate, (vi) polyisoprenoids such as natural rubber or Gutta percha, (vii) polyphenols such as lignin or humic acids, and (viii) polythioesters such as poly (3HB-co-3MP), which were reported recently. Biopolymers have the capability of undergoing the full biodegradation process at accelerated rates, and hence, under the enzymatic action of microorganisms, the breaking of biopolymers takes place cleanly into simple molecules such as water or methane and carbon dioxide found in the environment in a defined period of time [15].

5.1.2 Origin of Chitin and Chitosan

In 1811, the chitin biopolymer was first found in mushrooms by Professor Henri Braconnot, director of the Botanical Gardens at the Academy of Sciences in Nancy, France. The first appearance of the English word "chitin" comes from the French word "chitine" in 1836, and these words were derived from the Latin word "chitôn," meaning mollusk [16]. In the 1970s, when fishing companies were prevented from dumping shells of crabs and lobsters, the discovery of importance of chitin happened. After cellulose, the second most abundant organic and renewable source available in nature is the chitin, which is generated annually as waste in large quantities [17]. Chitin $(C_8H_{13}O_5N)_n$, a naturally occurring long-chain high-molecular-weight polysaccharide composed of an N-acetylglucosamine unit in the α (1–4 linkage), is found throughout the natural world especially in the exoskeleton of shellfish such as lobster or crabs, shrimp, radulae of molluscs, beaks of cephalopods, including squid, cell wall of fungi and scales of fish and lissamphibians, which accounts for about 10% of global landings of aquatic products [18, 19].

The chitosan extraction process was done by the process of demineralization of shrimp waste through acid treatment and deproteinization by alkali treatment. As an alternative treatment in recent years, several enzymatic deproteinization processes have been introduced for alkali digestion to reduce chitin quality loss due to depolymerization of its chain and also to produce protein hydrolysate with a well-balanced amino acid composition [20]. In addition, a new method of chitin extraction has been introduced using proteolytic microorganisms or fungi or purified enzymes, which are a low-cost technology and eco-friendly. Since the commercially available chitin powder is not soluble in any solvent but precipitates immediately, it has low workability and hence most of the chitin biopolymer is thrown away as commercial waste even though it is produced at a rate of 10^{10} to 10^{11} tons per year in nature [21]. Chitin's poor solubility is the major limiting factor in its utilization and the investigation of its properties and its structure. Despite these limitations, various applications of chitin and modified chitins have been reported in the literature. Wheat seeds treated by chitin biopolymer have growth-accelerating and growth enhancing effects. Significant reductions in the root-knot worm infestations and suppression of fungal pathogens resulted mainly from the chitinous additions to potting mixtures/soils [22].

Globally, including India, considerable chitin/chitosan research is in progress to tailor and impart required functionalities to maximize its utility. In fungi, the chitin polysaccharide involves cross-linkages to glucan polymers to create a meshed hyphal wall [23, 24] and due to the involvement of other polymers, such as glucans, the chitin content of fungal cell walls ranges from 22–40% [25]. Chitosan is produced between 3800 and 20,000 Daltons and is the structural element in the exoskeleton of crustaceans and cell wall of fungi [26]. Commercially, chitin and chitosan are now produced in Australia, Norway, India, Poland, Japan, and the United States.

Chitin and chitosan can undergo certain reactions such as etherification, esterification, and cross-linking. Chitin and its derivatives have exhibited high potential in a wide variety of fields, including pharmaceutical, cosmetics, bio-related science and technology, and environmental protection [27]. In several works, the production of structural biopolymers chitin and chitosan occupies an important place due to certain issues of intensification of processes [28]. Chitin and chitosan form the basis for many potential products used in fields such as textiles, photography, medicine, agriculture, and food processing. Research and development work on chitin and chitosan has witnessed intense activities in many parts of the world, since these biopolymeric materials are of great futuristic potential with immense possibilities for structural modifications to impart desired properties and functions [29].

5.1.3 Modifications of Chitin and Chitosan

One of the possible breakthroughs in chitin chemistry and technology is that the chemical modification of chitin is done more and more actively to explore highly sophisticated functions [30]. In recent years, in order to improve the solubility and the extension of applications, the chemical modification of chitosan and chitin has attracted growing attention [31, 32]. In spite of potential applications of chitin and chitosan, certain efficient appropriate modifications were established in chitin and chitosan to explore fully their high potential [33]. Chitin and chitosan are the versatile promising biomaterials of biodegradable nature. The presence of more number of reactive acetamido, hydroxyl, and amino side groups offers possibilities for chemical modifications leading to the production of a large variety of useful derivatives that are commercially available or can be made available via graft reactions and ionic interactions. The chemical modification of chitin and chitosan would not alter the fundamental skeleton of chitin and chitosan resulting in the generation of new biofunctional compounds. Depending on the nature of the group introduced, it would keep the unique physicochemical and biochemical properties [34, 35]. Without disturbing the overall effects of chitosan, considerable chances of chemical modification of this versatile biopolymer chitosan arise due to the availability of more number of functional groups such as primary amino group ((C-2) per-repeat unit), primary, and secondary free hydroxyl groups (two hydroxyl groups (C-3, C-6) [36]. New derivatives having promising biological activities and physiochemical properties have been developed from the chemical modification of chitosan.

A variety of chitosan derivatives have been synthesized to improve its biological properties, to advance its solubility in physiological media, and to widen its applications. Some important examples of modified chitosan that hold prominent place in research are quarternized chitosan, acyl chitosan, phosphorylated chitosan, N-alkyl chitosan, carboxy alkyl chitosan, thiolated chitosan, and sulfated chitosan. The modified quarternized chitosan has proved to be effective for the preparation of new antimicrobial drugs with high potential, since they possess enhanced antibacterial activity. The interaction of positive charge density of quaternized chitosan with the negatively charged cell surface of bacteria leads to the leakage of proteinaceous and other intracellular constituents [37]. The hydrophobic biocompatible and biodegradable chitosan membrane prepared by the modification of chitosan with butanal, hexanal, octanal, or decanal aldehydes, which can replace Nafion, plays an important role for electrode coatings utilized especially in both fuel applications and sensors [38, 39]. Due to its excellent biocompatibility and especially unique chemical, physical, and biological properties, N-carboxy methylated chitosan has a wide range of biomedical applications such as wound dressings, enhancement of permeation, and adsorption of low-molecular-weight heparin, an anionic polysaccharide across intestinal epithelia, antifungal, artificial bone and skin, apoptosis inhibitor, antibacterial, antioxidant, apoptosis inhibitor and blood anticoagulant [40–42]. When compared with chemical modification, chitin and chitosan modification by enzymes is interesting owing to its specificity and environmental advantages. The enzymatic grafting of phenolic compounds onto chitosan to confer water solubility under basic conditions has been reported by Kumar and coworkers [43]. Yoo and coworkers reported the selective modification of C-6 primary alcohol on chitosan by sequential TEMPO (2,2,6,6-tetramethyl-1-piperidine oxoammonium ion)-mediated oxidation, and the changes observed in water solubility at different levels of oxidation were monitored (from 25 to 100%) [44].

5.2 Characteristics of Chitin, Chitosan, and Derivatives

The biomaterials that are developed using natural polymers are an important and promising topic of research. Due to a large number of characteristics, including the presence of reactive groups, their polyelectrolyte and cationic nature, bacteriostatic and fungistatic influences, and high adsorption capacities [45–48], both chitin and chitosan biopolymer act as potential biomaterials for a variety of applications such as tissue engineering, drug delivery, wastewater treatment, food, and crop production. Several studies have found acylated chitosan to be a very interesting derivative of chitosan utilized in biomedical applications. Chitosan and its derivatives such as trimethyl chitosan have been used in non-viral gene delivery. Nowadays, the rapidly developing field of tissue engineering encompasses all the major scientific disciplines, including materials science, polymer chemistry, biology, and medicine. The main requirements for the elaboration of tissue engineering scaffolds stipulate that the prepared scaffold should be biodegradable, so that the cured tissue will be able to replace the biomaterial and possess surface properties that will promote cell attachment, differentiation, and proliferation.

Another important requirement is that the scaffold should not induce acute or chronic response. Based on the creation of suitable mechanical properties for handling and to mimic the damaged tissue, finally it can be manufactured into a variety of shapes [49, 50]. Due to their unique structural, physico-chemical and functional properties, chitin and chitosan derivatives appear to be suitable candidates for the wastewater treatment, preparation of dressing materials, and scaffolds for tissue regeneration [51]. Due to the various aforementioned remarkable properties, chitosan and its derivatives appear to be a relevant candidate that could substitute for missing or damaged tissue and organ [52] and allow cell attachment and proliferation [53] provided that 3D-scaffolds might be produced.

5.2.1 Physicochemical Properties of Chitin, Chitosan, and Their Derivatives

The chitin biopolymer is a off-white nitrogeneous polysaccharide of hard and inelastic nature [54]. The important characteristic of the biopolymer is its ability to degrade by natural enzymes with their immunogenic behavior. Chitin and chitosan are highly basic polysaccharides that have several properties such as solubility in various media, polyelectrolyte behavior, metal chelating, solution, viscosity, polyoxysalt formation, film-forming ability, and optical and structural characteristics [55]. The chemical properties of chitosan are mainly due to the presence of linear polyamine and reactive amino groups, which have the tendency to chelate with many transitional metal ions. Some factors during processing such as the degree of deacetylation, molecular weight, concentration of solution, ionic strength, pH, and temperature affect the production of chitosan and its properties giving rise to the definition of "intrinsic viscosity" of chitosan, which is a function of the degree of ionization as well as ion strength.

When compared to chitin, one of the greatest advantages of chitosan is that it can be easily fabricated without the use of harsh chemicals into many forms such as fibers, microgels, films, and nanogels under mild conditions since it is soluble in simple dilute acids or aqueous solutions. Crystallinity is the hydrogen bonding formed between the arranged strands of chitin; the higher the crystallinity, the more stability for chitin molecules, and the degree of crystallinity is a function of the degree of deacetylation [56]. For chitin, the degree of deacetylation usually ranges 5–15% and for chitosan it is in the range 70–95% [57]. The molecular weight of native chitin is usually larger than 1 million Daltons, while commercial chitosan products have the molecular weight range of 100,000–1,200,000 Daltons [58]. The average molecular weight of chitin ranges from 1.03×10^6 to 2.5×10^6 Dalton, but the conversion of chitin to chitosan by deacetylation reduces it to 1×10^5 to 5×10^5 [59]. Chitin is hydrophobic (water insoluble) as well as not soluble in most organic solvents; on the other hand, it dissolves only in solvents such as *N*,*N*-dimethylacetamide, hexafluoroacetone, and hexafluoro-2-propanol. In contrast to chitin, chitosan is soluble in dilute organic acids at low pH [60].

Since the presence of amino and hydroxyl groups which they easily substituted with other groups, both chitin and chitosan are highly reactive and can be easily processed into gels, beads, powders, fibers, membranes, cotton, flakes, sponges, colloids, films, and spins and also were found to be remarkably stable in concentrated alkaline solutions even at high temperatures. The solubility properties of chitin and chitosan highly depend on the number of free amino groups on the molecules. The amino groups in the chitosan backbone become protonated under acidic pH and the polymers dissolve in polar solvents. The chitosan molecule lose their positive charge under neutral and alkaline pH resulting in precipitation. These properties can be exploited to prepare chitosan matrices in various forms such as capsules, powders, hydrogels, micro- and macrofiber, membrane, nanofiber, macro- and micro-beads, and scaffolds.

The physicochemical properties and functional characteristics of five commercially available chitins and chitosan were investigated by Young In Cho and coworkers. The obtained results indicate that the physicochemical characteristics (nitrogen, ash, degree of deacetylation, bulk density, and viscosity) differed with products. Functionally the dye-binding capacity differed depending on the products, although the average binding capacity (63%) of chitosan was higher than that (54%) of chitins. For chitins, the water-binding capacity ranged from 381% to 673% and for chitosan it varied from 458% to 805%. Except for one product (563%), the fat-binding capacities of chitins were mostly similar (316–320%), whereas for chitosan the fat-binding capacity showed dissimilar binding capacities from 314 to 535%. Significant correlations were observed between fat-binding capacity and viscosity ($r = 0.72$, $P < 0.05$) and between water-binding capacity and bulk density ($r = -0.89$, $P < 0.01$).

Both water and fat-binding capacity of chitosan products were significantly correlated positively with ash ($r = 0.81$, 0.80) and negatively with bulk density ($r = -0.98$, −0.95) [61]. Fai and coworkers reported about the physico-chemical characteristics and functional properties of chitin and chitosan produced by *Mucor circinelloides* using yam bean as substrate. The observed results demonstrated the highest levels of chitosan (64 mg/g) and chitin (500 mg/g) were produced at 48 and 72 h, respectively [62].

5.2.2 Biological Properties of Chitin, Chitosan, and Their Derivatives

Chitin and chitosan possess very interesting biological properties and therefore have been mainly used in many applications of medical and pharmaceutical fields [63–65]. Both chitin and chitosan possess eco-friendly, non-toxic, and biodegradable nature and also have biological properties such as hemostatic, anticholesteremic, biocompatible, citocompatible, antimicrobial, antioxidant, fungistatic, spermicidal, antitumor, anti-inflammatory, analgesic, and mucoadhesion [66]. Chitins and chitosan derivatives synthesized in a variety of fashions can be used to prove molecular hypothesis for the biological activity. These biological properties accelerate the formation of osteoblast, which is responsible for bone formation. Table 5.1 represents the relationship between some chitin and chitosan biological properties and their physicochemical characteristics [67].

Table 5.1 Relationship between Chitin and Chitosan biological properties and their characteristics

Property	Characteristic
Hemostatic	Degree of deacetylation, length of chains (molecular weight)
Biodegradability	Degree of deacetylation, molecular weight, distribution of acetyl groups
Anticholestrolemic	Degree of deacetylation, molecular weight, viscosity
Mucoadhesison	Degree of deacetylation, molecular weight (only chitosan)
Biocompatibility	Degree of deacetylation
Antimicrobial	Molecular weight
Adsorption enhancer	Degree of deacetylation (only chitosan)
Analgesic	Degree of deacetylation
Antioxidant	Degree of deacetylation, molecular weight

The biological properties of these compounds were found to be strongly dependent on their solubility in water and other commonly used solvents. Above pH 7, the chitosan is normally insoluble in aqueous solutions in its crystalline form; however, in dilute acids, the protonated free amino groups facilitate the solubility of the molecule. Majority of the biological properties of chitosan were related to its cationic behavior. The pKa of primary amino groups depends closely on degree of acetylation, and hence, the solubility of chitosan is also dependent on the degree of acetylation. In addition, the biological properties of chitin/chitosan samples can also be related to the crystallinity, polydispersity (MW/MN), and distribution of GlcNAc and GlcN units along the polymeric chain described by the pattern of acetylation (PA) [68].

Since the degradation is essential in many small and large molecule release applications, the understanding and control of the degradation rate of chitin and chitosan-based devices is of great interest and several studies reported that the length of the chains (MW) also affects the degradation rate [69–71]. Moreover, the biodegradability process gets affected by the distribution of acetyl groups since the absence of acetyl groups or their homogeneous distribution (random rather than block) results in very low rates of enzymatic degradation [72]. The authors reported that the differences in degradation processes were mainly observed due to the differences in deacetylation conditions, which influence the viscosity of chitosan by changing the inter- or intra-molecular repulsion forces [73–75].

Depending on some of the sample characteristics such as the natural source, method of preparation, MW, and degree of deacetylation, both chitin and chitosan biopolymers show very good compatibility. When compared to chitin biopolymer, chitosan biopolymer presents higher cytocompatibility in vitro. The chitosan cytocompatibility has been proved in vitro with myocardial, endothelial, and epithelial cells, fibroblast, hepatocytes, condrocytes, and keratinocytes [76]. Different hemostatic mechanisms were found to be followed for the solid-state chitosan and chitosan acetic acid physiological saline solution. Results indicated that the solid-state chitosan with a high DD bound more platelets and hence was found to be more hemostatic [77].

The analgesic effect of chitin and chitosan on inflammatory pain due to intraperitoneal administration of acetic acid has been reported by Okamoto and coworkers [78]. The reported results herein indicate that the chitin shows a lesser effect when compared to chitosan and this difference was mainly due to the different action mechanisms of the two polymers. Due to the polycationic nature of chitosan, the reduction in pH occurs by the protonation of primary amino groups in chitosan and this reduction in the pH is the main cause of the analgesic effect. In both in vitro and in vivo analyses, chitosan oligomers possess antitumor activities, and this was reported by Jeon and Kim. Partially deacetylated chitin as well as chitin with a carboxymethyl group has also been effective in demoting tumor progression [79].

Oxidative damage and the risk for age-related disorders can be minimized by antioxidative nutraceuticals, such as tocopherols, ascorbic acid, caroteinoids, polyphenols, and chitosan. Literature survey reveals that both chitin and chitosan biopolymer possess antimicrobial activity. The most important characteristics for a polymer to act as an ideal antimicrobial were the simplicity to synthesize, high stability for a vast period of usage and storage at the temperature of its intended application, easy solubility in water or neutral media, non-toxic, regeneration upon loss of activity, and biocidal to a broad spectrum of pathogenic microorganisms in brief times of contact. The antimicrobial property of chitosan depends on various factors, which include the type of chitosan used (whether in native or modified form), its degree of polymerization, the host, the chemical and/or nutrient composition of the substrates, and environmental conditions [80]. Some studies reported that the oligomeric chitosan exhibits excellent antifungal activity than larger units.

The antimicrobial activity of chitosan depends not only on the external conditions (target microorganism, nature of the medium, pH temperature, etc.) but also on different intrinsic factors such as its molecular weight and degree of polymerization and deacetylation [81]. Two theories have been proposed for explaining the antimicrobial mechanism of chitosan. Theory one states that the polycationic nature of chitosan is the reason for its antimicrobial nature [82] to interact with the negatively charged microbial cell membranes, which results in the disruption of the cytoplasmic membrane and, ultimately, leakage of intracellular constituents [83]. According to the other theory, chitosan oligosaccharides easily permeate into the nucleus of eukaryotic cell and interfere with the transcription of RNA and the synthesis of proteins [84, 85]. Tran Minh Quynh reported in his study that the irradiation of chitosan through gamma radiation modifies the intrinsic viscosity and the number average molecular weight, which in turn improves its antibacterial and antifungal activities. This irradiated chitosan can be applied as antimicrobial coating films for the preservation of fresh fruits [86].

Chitinolytic microbes produce extracellular chitinase enzymes to degrade chitin-rich tissues of other organisms. Chitinase enzyme hydrolyses the chitin to produce chito oligomers for the control of plant pathogenic fungi, a major problem for agricultural food production. Saima and coworkers developed novel mesophilic bacterial strains using *A. hydrophila HS4* and *A. punctata HS6* to produce chitinase enzyme, which showed excellent defense against pathogenic fungi [87]. Other studies showed that the chitinase enzyme produced from *Streptomyces rubiginosus* shows maximum activity after 72 h of incubation period after which the activity decreases gradually and it acts against plant pathogenic fungi. The addition of chitin to the soil promotes the growth of antagonistic microbes. For example, the bacterium *Bacillus subtilis*, a pathogen of fungi, is used as biopesticide in agriculture [88]. *Bacillus subtilis* along with chitosan was also used against the powdery mildew in strawberry [89]. The pathogen *Streptomyces scabies* causes scab disease on potato tubers and it was controlled by the addition of chitin.

5.3 Recent Research in Applications of Chitin and Chitosan in Agriculture

Chitin and chitosan biopolymers, generally called a "defense booster," play especially a vital role in agriculture, including boosting defense mechanism of plants against invading microorganisms. The protection of crops from pests, pathogens, and physiological disorders has been repeatedly carried out by chitin and its derivatives [90]. Natural chitosan elicitor solution for agriculture and horticultural uses was granted an amended label for foliar and irrigation applications by the EPA in 2009. The US Forest Service tested chitosan as an eco-friendly biopesticide to prearm pine trees to defend themselves against mountain pine beetles. Both chitin and chitosan have been explored for many agricultural uses due to their antiviral, antibacterial, and antifungal properties. A few other applications of chitin and chitosan in agriculture are presented below.

5.3.1 In Preservation of Agricultural Commodities

During the past 100 years, the use of chemical fertilizers and pesticides in agriculture raised productivity significantly, but nowadays there is a growing international concern over the indiscriminate use of synthetic chemical agents on crops, since it introduces the harmful effects on human health; hence, chemical fertilizers and pesticides led to reduced biological diversity and degraded natural and agricultural systems [91]. In order to overcome the harmful effects and also to reduce the use of synthetic chemical agents, there is a worldwide trend to explore new alternatives. Certain alternative methods are necessary to face these challenges of reducing the environmental impact of the activity without affecting agricultural productivity and with economic benefits for farmers. Among various alternatives, chitin and chitosan biopolymers have become a promising alternative treatment utilized for crop protection due to their antimicrobial activity [92].

Chitosan has been proven to control numerous pre- and post-harvest diseases on various horticultural commodities. Recently new alternatives were explored worldwide to control postharvest pathogenic diseases and giving priority to methods that reduce disease incidence, avoiding negative and side effects on human health as a result of the excessive application of synthetic fungicides [93]. It is important to enumerate and identify molds that occur in food commodities for both quality assurance and control operations in food processing plants [94]. Chemical determination of chitin, a constituent of fungal cell walls, has an advantage that it will reflect total mycelium based on chitin content [95]. Bishop and coworkers used chitin to further evaluate the detection of mold in tomato products, ketchup, paste, and puree. Variations were observed in chitin content among different fungal species, depending upon cultural age, and growth conditions, and values ranged from 5.7 to 43 mg of glucosamine per mg dry weight [96].

In recent years, chitosan coatings have emerged as an ideal alternative to chemically synthesized pesticides. They have been extensively used in agriculture as coating materials by dipping or spraying of chitosan on to the fruits, seeds and vegetables such as capsicum, carrot, cucumber, grapes, green peas, potato, radish, and tomatoes for increased host defense. They can also help protect the safety of edible products. The chitosan coating stimulates the plant to produce systematic antibodies, which induce repellent effects to the crop pests [97]. Bautista-Banos and coworkers reported in their work that anthracnose disease of papaya fruit was better controlled when the fruits were dipped in a chitosan solution of 1.5% before rather than after artificially inoculating fruit [98]. Similarly, in another study, mango fruit was dipped in chitosan solution, and it had lower reducing sugars than control fruit [99, 100]. Indicating a higher reduction of the fruit metabolism in the dipped than in the non-dipped fruit. The content of ascorbic acid was also evaluated in chitosan-treated mangoes and peaches [101]. In those studies, the content of this vitamin in the treated mango gradually decreased during the storage period and was lower than in untreated fruit. However, for peaches, the content of ascorbic acid was higher in chitosan-treated fruit when compared with untreated or in prochloraz-treated fruit after a 12-day storage period.

Pomegranate fruits coated with chitosan had delayed the decrease in total soluble solids (TSS) and titratable acidity (TA) percent [102]. Abdel-Mawgoud and coworkers reported that the chitosan foliar application on strawberry crops increases plant height and number of leaves, augments leaf fresh and dry weights, and enhances the number and weight of seedlings [103]. Chitosan was most effective in controlling postharvest decay when it was dissolved with organic solvents such as ethanol for prolonged cold storage of table grapes, and this combination of reduced doses of chitosan and ethanol improves the synergistic effects in the control of gray mold of table grapes [104]. Chitosan treatment has been reported to influence antioxidant enzyme activities in the tissues of fruits and vegetables. Compared to untreated strawberries, those treated with chitosan maintained higher level of antioxidant enzyme activities, such as catalase, glutathione-peroxidase, guaiacol peroxidase, dehydroascorbate reductase, and monohydroascorbate reductase [105]. Compared to the tissues of uncoated fruit, higher activities of superoxide dismutase, catalase, and peroxidase were reported after chitosan application to pears [106].

Eva-Guadalupe Lizárraga Paulin and coworkers studied the effect of chitosan and hydrogen peroxide seed coatings and sprinkling on the seedling of two different varieties of maize. He reported that the application of chitosan and hydrogen peroxide to maize plants enhances the nutritional quantity in the germination stage [107]. Similarly, peanut seeds soaked in chitosan were reported to exhibit an increased rate of germination and energy, lipase activity and gibberellic acid and indole acetic acid levels [108]. Ziani and coworkers found that chitosan coating significantly differed on the increasing artichoke seed germination and decreased the number of type of fungi and increased plant growth. In addition, chitosan with lower MW gave better results from both microbial and germination process [109]. Hirano and coworkers reported that the coating of seeds with depolymerized chitosan or its oligosaccharides typically increased the chitinase activity in seedlings by 30–50%, unless the seeds had a hard cuticle. Low molecular weight of chitosan (5 kDa) induced the accumulation of phytoalexins in the plant tissue, decreased the total content and changed the composition of free sterols producing adverse effects on infesters, activated chitinase, beta-glucanase, and lipoxygenases, stimulating the generation of reactive oxygen species. Chitosan protection by exclusion occurs with soybean seed treatments. In this case, the major advantage was protection from insects such as agarotis, ypsilon, soybean pod borer, and soybean aphids [110].

5.3.2 In Plant Resistance to Pathogens and Defense Mechanisms

In order to induce plant defense mechanisms, a wide variety of elicitors have been employed to alter cell metabolism, which will enhance the production of secondary metabolites in plants and plant cell cultures [111–113]. Chitosan, a deacylated derivative of chitin, is known to promote plant defense and growth, but the mechanisms by which it is able to do so are still unclear. Chitin can be hydrolyzed into oligomers and monomers by acid hydrolysis and these oligo-chitins have been proposed as anti-microbial agents, promoters of plant growth, elicitors of plant resistance, enhancers of the immune response, and agents against malignant growth [114]. The chitosan biopolymer's effects on plant response were first characterized as elicitor and it activated the plant defensive genes through the octadecanoid pathway [115]. It was proved in several studies that chitosan is effective in inducing disease resistance in plants, with pathogen and plant cultivar specificity [116, 117].

Certain studies which have been performed by utilizing the natural defense mechanisms reveal that plants fight off microbial invasions through induced responses involving signal molecules [118, 119]. These signal molecules, or elicitors, generally enhance non-specific plant resistance due to their effect of mimicking a pathogenic attack, thereby priming plant defenses before infection [120]. A diverse range of elicitor molecules have been reported, including oligosaccharides, polysaccharides, lipids, glycoproteins, peptides, and proteins [121]. Soil amendment with chitosan has been extensively reported in many works on fungal diseases. These studies clearly show chitosan to be fungistatic against both biotrophic and necrotrophic pathogens. For soilborne diseases, a major issue is the suppressiveness of soil toward pathogens that may occur in the soil [122]. In this context, it has long been suspected that suppressiveness can be enhanced by adding biopolymers such as chitin and derivatives. For instance, soil treatment with chitin and/or chitosan from shrimp waste has been shown to temporarily increase root growth (e.g., of tomato) and decrease the rate of infection of plant roots by nematodes [123–125]. Although not definitely proven in all cases, the mechanisms behind this suppressiveness enhancement most often relate to a change in the structure and/or activity of the microbiota in soil, which thus confers suppression of plant pathogens [126].

Presumably, chitinolytic microorganisms, which are capable of hydrolyzing the chitinous hyphae of pathogenic fungi, increased their numbers and/or activities in response to the chitin addition. Alternatively, secondary responders to the added chitin confer pathogen suppression. The triggered defense mechanisms of chitosan in plants are discussed below. Plants treated with chitin and chitosan produce chitinase, which breaks down the chain of chitin and chitosan into a more soluble form [127]. Chitosan enhances the innate immunity by stimulating hydrogen peroxide (H_2O_2) production in rice [128, 129], stimulates drought resistance of coffee [130], induces a stress response by nitric oxide (NO) pathways in tobacco [131, 132], helps in the synthesis of phytoalexin [133], and influences pathways involving jasmonic acid-ethylene (JA/ET) signaling marker in oilseed rape [134].

5.3.3 In Biostimulation of Plant Growth and Its Protection

Chitin and chitosan are naturally occurring compounds that have potential in agriculture with regard to controlling plant diseases. These molecules were shown to inhibit fungal growth and development. They were reported to be active against viruses, bacteria, and other pests. In recent years, many researchers reported that the chitosan biopolymer has enhanced the plant growth and development [135–137]. Results reveal that the application of chitosan has increased the key enzymes activities of nitrogen metabolism (nitrate reductase, glutamine synthetase, and protease) and improved the transportation of nitrogen (N) in the functional leaves, and this resulted in enhanced plant growth and development. Fragments from chitin and chitosan are known to have eliciting activities leading to a variety of defense responses in host plants in response to microbial infections, including the accumulation of phytoalexins, pathogen-related (PR) proteins and proteinase inhibitors, lignin synthesis, and callose formation. Based on these and other properties that help strengthen host plant defenses, there has been growing interest in the use of chitin and chitosan in agricultural systems to reduce the negative impact of diseases on the yield and quality of crops [138].

The chitosan biopolymer has been proven to protect the safety of edible products, to induce abiotic and biotic stress tolerance in various horticultural commodities, and to stimulate plant growth. It has been used to protect plants in space, as well, exemplified by NASA's experiment to protect adzuki beans grown aboard the space shuttle and Mir space station in 1997. NASA results revealed chitosan induces increased growth (biomass) and pathogen resistance due to elevated levels of beta 1,3-glucanase enzymes within plant cells. NASA confirmed chitosan elicits the same effect in plants on earth. Several reports suggest the effect of irradiated chitosan as a postharvest preservative edible coating on oranges, mango, and bell pepper. Irradiated chitosan coating on these fruits reduces the weight loss and respiration rate during storage, delays and minimizes the changes in ascorbic acid contents, titratable acidity, and preserves the taste and quality of fruits, as compared to untreated fruits.

Mariana Silvia Cretoiu and coworkers conducted a long-term experiment for 8 years in an experimental field, Vredepeel, the Netherlands. The data revealed that chitin addition to soil had raised the suppression of soil, in particular toward *Verticillium dahliae*, 9 months after the (second) treatment, extending to 2 years following treatment. At the functional gene level, the abundance of specific (family-18 glycoside hydrolase) chitinase genes carried by the soil bacteria also revealed upshifts as a result of the added chitin. The effects of chitin noted for the *Oxalobacteraceae* were specifically related to significant upshifts in the abundances of the species *Duganella violaceinigra* and *Massilia plicata*. Thus Mariana Silvia Cretoiu and coworkers in their study concluded that chitin acts against borne pathogens and stimulate the plant growth [139].

Chitin promotes the growth of plants independently by promoting rhizobacteria, which causes increased plant growth and is sometimes associated with the biological control of plant pathogens, nutrient cycling, and seedling establishment and enhances the growth even in sterile conditions. This is backed by other studies, including tissue-cultured grapes and the growth of the medicinal herb phyla dulcis being cultivated in liquid bioreactors. Similar observation was made with chitosan, as it was effective even at very low concentration of 10 mg L^{-1} and stimulates the growth of orchid.

5.3.4 In Enhancing Food Production

Nowadays food safety is one of the critical issues, and hence, before being consumed by people, the agricultural product such as fruits and vegetables should be preserved in a good condition. The fungal infection on storage is one of the main reasons for the decrease of food quality and safety. Pre-harvesting, post-harvesting, handling, and transportation lead to wounds, and through these wounds the fungi spores get deposited resulting in the decay of fruits. The excessive amount of fungicides was utilized to control the pre- and post-harvest diseases in fruits; since it leads to an increase in the resistance level of fungi but in addition, it also induces the negative impact on human health.

As an alternative to fungicides, chitosan, the defense booster, has been used since the permeability of chitosan coating could be showed by resistance to the gases such as O_2, CO_2 and the respiration level. The poly-D-glucosamine units present in chitosan binds to fungi receptors and mimick the attack by spores and give signals to plants which bring out various genetic and biological responses. The level of inhibition of fungi mainly depends on the concentration of chitosan. Owing to its polycationic nature and polymer chain length, chitosan possesses enhanced antifungal activity; hence, it is recommended to be used as biopesticide for post-harvest plants. In post-harvest storage, the chitin-based treatments show promise as alternatives to the use of synthetic pesticides on fresh products [140, 141]. The agricultural applications of chitosan can reduce environmental stress due to increased yields, strengthened seed vitality, improved stand quality, and reduced fruit decay of vegetables, fruits, and citrus crops.

The effect of foliar application of chitosan as a growth promoter on morphological, growth, biochemical, and yield attributes and fruit yield of okra cv (Hibiscus esculentus L known as lady's finger) was investigated by Mondal and coworkers [142]. Five different levels of chitosan concentrations viz., 0 (control), 50, 75, 100, and 125 ppm were sprayed three times at 25, 40, and 55 days after sowing. Results revealed that most of the morphological (plant height, leaf number plant-1), growth (total dry mass plant-1, absolute growth rate, relative growth rate), biochemical parameters (nitrate reductase and photosynthesis), and yield attributes (number of fruits plant-1 and fruit size) were increased with increasing concentration of chitosan until 25 ppm, and this resulted in the highest fruit yield in okra (27.9% yield increased over the control). However, the increment of plant parameters as well as fruit yield was not significant from 100 ppm of chitosan and therefore the foliar application of chitosan at 100 or 125 ppm may be used at the early growth stage to achieve a maximum fruit yield in okra.

Sivakumar and coworkers reported that by retarding the ripening process and color development, the papaya fruit quality has been improved by chitosan treatment and in addition it also controls the rots on stored papaya by controlling anthracnose (*Colletotrichum gloeosporioides*). Due to chitosan treatment, the fruits were firmer and lost less weight compared with control treatments [143]. Suchada Boonlertnirun and coworkers reported about the utilization of chitosan in rice production. The results showed that the application of chitosan by seed soaking and soil application four times throughout the cropping season significantly increased rice yield over the other treatments [144]. The application of chitosan oligosaccharides as a soil conditioner with high and low nitrogen content of the soil was studied by

Utsunomiya and Kinai [145]. Their study revealed that under high nitrogen content, the soil conditioner COS increased the growth in the number of flowers, fruits, fruit weight, and juice production. The low-nitrogen condition decreased the flower formation in the first growing season and had little effect on flowering in the next season.

The effect of chitin as a feed additive on the growth of bifidobacteria in the guts of chickens has been reported by Austin and coworkers [146]. The reported results indicate that the addition of chitin increased the growth of bifidobacteria, which are important as they inhibit the growth of other types of microorganisms, and in addition the bifidobacteria also generate the lactase required for the digestion of milk lactose which may be of significance for humans and animals with lactose intolerance [147]. The application of chitosan at early growth stages by Chibu reveals that the plant growth and development has been increased thereby resulting in increased seed yield in rice and soybean [148]. Similar results were also reported by Boonlertnirum and coworkers [149] in rice and by Rehim and coworkers [150] in maize and bean.

5.3.5 In Enhancing Crop Growth

Chitin and chitosan have been potentially used as a plant growth enhancer. The host defense to defeat fungal infection and pathogens was boosted up by both biopolymers. The defense response is increased to conquer microbial infection, including phytoalexins and proteinase inhibitors, lignin synthesis, accumulation of pathogen-related proteins, and callose formation. A number of modes of action such as the direct antibiosis and the induction of plant defenses have been identified for the beneficial effects of chitin-based treatment on crops. The action in stimulating beneficial microbes has proved particularly impressive, with chitin/chitosan amplifying the effect of beneficial microbes in remediating soil pollutants, controlling pathogens, and promoting plant growth. The subsequent responses of plants and microbes by chitin addition have led to improvements in disease control, plant growth, and ultimately improved crop yield and quality.

The enhancement of crop growth by chitosan treatment on plants was due to the inhibition of the growth of different pathogens by their activation of protective mechanism in plant tissues; hence, chitosan acts as an alternative to chemical fungicides. The promotion of plant growth by chitosan treatment leads to improved yields and plant health in numerous crops and fruits. The spraying of chitosan to plant seeds enhances the germination index, the mean germination time is reduced, and shoot height, root length, and seedling vigor are increased [151]. In addition, the seed quality, energy of germination, germination percentage and lipase activity have been enhanced by the chitosan-mediated seed treatment, and these treasures can be used in many ways to preserve and increase the crop yield, quality of fruits and vegetables. For increased host defense, chitosan has been widely applied in a range of fruits and vegetables including cucumber, capsicum, carrot, grapes, green peas, potato, radish, and tomatoes. When applied to foliage or soil, the biocontrol mode of action of chitosan elicits natural innate defense responses within the plant to resist pathogens, insects, and soilborne diseases.

Chitosan stimulates nutrient uptake, increases germination and sprouting, boosts plant vigor, promotes plant growth, and increases photosynthesis. When used as seed treatment or seed coating on cotton, corn, seed potatoes, soybeans, sugar beets, tomatoes, wheat, and many other seeds, it elicits an innate immunity response in developing roots which destroys parasitic cyst nematodes without harming beneficial nematodes and organisms. Chitin-based treatments showed numerous positive responses in field-grown crops, and in addition, numerous studies have identified positive effects in other sectors of agriculture and horticulture, too. The growth of micropropagated crops has been shown to increase by the inclusion of chitosan to the growth medium [152–155]. The in vivo studies showed that irradiated chitosan improves crop yields and suppress diseases due to virus, bacteria, and fungi [156]. A low-molecular-weight chitosan (MW 14 k Dalton) called "Fitosan" was prepared by irradiating chitosan using gamma rays at a dose of 75 kGy and prepared from the deacetylation of chitin using a NaOH solution at 90°C for 8 h. It was successfully tested in the field for several plants such as chili, potato, soybean, and rice to improve crop yields and to suppress the diseases.

Wu and co-workers developed a double-coated slow-release NPK compound fertilizer prepared using chitosan as inner coating and cross-linked poly(acrylic acid)/diatomite—containing urea as the outer coating. This fertilizer acts as a superabsorbent polymer possessing excellent water retention capacity. It was noticed that the product showed a slow controlled release of the nutrients without any detrimental impact on the soil and the nutrients released did not exceed 75% on the 30th day [157]. A preliminary study was carried out by Corradini and coworkers to incorporate the chitosan nanoparticle in the NPK fertilizer using urea, calcium phosphate, and potassium chloride, respectively, for the nitrogen, phosphorus, and potassium contents. He concluded from his study that the stability of the chitosan nanoparticle suspension was higher with the addition of nitrogen and potassium than with the addition of phosphorus, due to the higher anion charge from the calcium phosphate than the anion charges from the potassium chloride and urea [158]. Hamid and coworkers showed that the chitosan-based urea fertilizer prepared using chitosan, urea, and bentonite clay with the clay as backfiller material behaves as a very good promising material in crop protection. From his study, he proved that the water absorption and weight loss percentage of the produced chitosan-based urea fertilizers were found to be directly proportional to the increasing chitosan contents; this might be due to the increase in hydrophilic properties. This fertilizer also possesses the property of slow controlled release of urea [159].

Santos and coworkers prepared potassium-containing, controlled-release biodegradable material based on chitosan and montmorillonite clay layered silicate recently and monitored the release of fertilizer using time-domain reflectometry (TDR). Chitosan-Montmorillonite (ChMMT33)-containing potassium shows two specific periods of fertilizer release: the first one lasted approximately three days and was assigned to the external fertilizer on the microspheres. The second was assigned to the internal fertilizer [160]. Narayanan and Dhamodran prepared a new super water-absorbing, material called CHEDUR using chitosan, EDTA, and urea. CHEDUR is very promising for its applications that demand the rapid absorption and slow release of urea, water and other metal ions that can be attached through the EDTA component [161]. Bita Roshanravan and coworkers prepared urea-intercalated kaolinite fertilizer using chitosan as binder, and then it was granulated to prepare nitrogen-based controlled-release properties (CRF). The optimum concentration to yield granule was 6% of chitosan. This research group correlated that there is direct relationship between granule size and compressive stress. Ammonia volatilization was carried out using the forced-draft technique for a period of 10 weeks. The results showed that the total amount of ammonia loss for conventional urea fertilizer and urea–kaolinite–chitosan granules was 68.63 and 56.75%, respectively, which confirm the suitability of fertilizer in agricultural crop protection [162].

5.3.6 In Enhancing the Production of Plant Metabolites

Secondary metabolites produced from plants play a crucial role in defense and adaptation to the environment. Some examples of secondary metabolites are terpenes, steroids, phenolics, and alkaloids, which exhibit a wide range of biological activities and have immense potential application in chemical-pharmaceutical industries as agrochemicals, pharmaceuticals, colors, flavors, fragrances, biopesticides, and food additives [163, 164]. As an alternative to classical approaches, the plant in vitro cultures represent an attractive and cost-effective for plant secondary metabolite (PSM) production.

Elicitation has been proved one of the most effective strategies utilized for the biotechnological production of plant secondary metabolite. Traditionally, according to their chemical nature, the elicitors have been classified in two types: abiotic or biotic. Based on origin, they can be classified as exogenous or endogenous, which notably include yeast extract, methyl jasmonate, salicylic acid and vanadyl sulfate [165]. Elicitation is a tool extensively used for enhancing secondary-metabolite yields. In order to induce phytoalexin accumulation in plant tissue, certain biopolymers such as chitin and chitosan were used as elicitors [166]. By antifungal responses, chitosan induces phytoalexins' secondary metabolite accumulation and enhances protection from further infections [167].

Md. Abdullahil Baque and coworkers reported about the effect of chitosan and pectin on the accumulation of anthraquinones

(AQ), phenolics, and flavonoids by adventitious root suspension cultures of *Morinda citrifolia* in shake flasks. The results show that the adventitious root cultures elicitor treated with various chitosan/pectin combinations or chitosan alone resulted in enhanced biosynthesis of secondary metabolite but strongly inhibited the root growth which might be due to the lethal effect of elicitor as evidenced by measurement of 36–79% cell death. The optimum elicitor concentration for enhancing metabolite biosynthesis was found at 0.2 mg/ml chitosan concentration, in which 103.16, 48.57, and 75.32 mg/g dry weight (DW) of AQ, phenolics, and flavonoids, respectively, were achieved [168].

The presence of micronutrients such as manganese, boron, copper, iron, chlorine, molybdenum, and zinc promotes optimum plant growth. Tao and coworkers synthesized chitosan modified with 1-naphthylacetic acid, which is an important plant growth hormone. They observed that the release of the 1-naphthylacetic acid was strongly dependent on pH and temperature and could continue for 55 days at pH 12 and 60°C. Despite this dependence, the formulation offers a potential for the slow release of plant growth hormones and enhances the plant metabolites [169].

5.4 Conclusion

This chapter summarizes the novel agricultural applications of chitin and chitosan. Multifunctional use of chitin and chitosan-based biomaterials has been proved to aid the plant growth and enhance food production. We expect that this chapter will provide insights on the use of these chitin and chitosan biomaterials for researchers working to discover new materials with new properties for valuable applications.

Acknowledgements

The authors are grateful to the authorities of DKM College for Women and Thiruvalluvar University, Vellore, Tamil Nadu, India, for the support. Thanks are also to the editor for the opportunity to review such an innovating field.

References

1. Malerba, M., and Cerana, R. (2016). Chitosan effects on plant systems, *Int. J. Mol. Sci.*, **17**, pp. 1–15.
2. Ekebafe, L. O., Ogbeifunand, D. E., and Okieimen, F. E. (2011). Polymer applications in agriculture, *Biokemistri.*, **23**, pp. 81–89.
3. Armentano, I., Bitinis, N., Fortunati, E., Mattioli, S., Rescignano, N., Verdejo, R., Lopez-Manchado, M., and Kenny, J. (2013). Multifunctional nanostructured PLA materials for packaging and tissue engineering, *Prog. Polym. Sci.*, **38**, pp. 1720–1747.
4. Akila, R. M. (2014). Fermentative production of fungal Chitosan, a versatile biopolymer (perspectives and its applications), *Adv. Appl. Sci. Res.*, **5**, pp. 157–170.
5. Johan, R. C., and Van der Marrel (2007). *Introduction to Biopolymer Physics*, 1st ed. (World Scientific, New Jersey).
6. Mohanty, A. K., Misra, M., Drzal, L. T., Selke, S. E., Harte, B. R., and Hinrichsen, G. (2005). *Natural Fibers, Biopolymers and Biocomposites* (CRC Press, Boca Raton).
7. Chandra, R., and Rustgi, R. (1998). Biodegradable polymers, *Prog. Polym. Sci.*, **23**, p. 1273.
8. Meyers, M. A., et al. (2008). Biological materials structure and mechanical properties, *Prog. Mater. Sci.*, **53**, p. 1.
9. Kumar, A., et al. (2007). Smart polymers: Physical forms & bioengineering applications, *Prog. Polymer Sci.*, **32**, p. 1205.
10. Ekiert, M., Mlyniec, A., and Uhl, T. (2015). The influence of degradation on the viscosity and molecular mass of poly(lactide acid) biopolymer, *Diagnostyka,* **16**, pp. 63–70.
11. John, M., and Thomas, S. (2012). Natural polymers: An overview, in: *Natural Polymers: Volume–1: Composites.* RSC Green Chemistry Series, Royal Society of Chemistry Publishing Company.
12. Mohanty, A. K., Misra, M., and Hinrichsen, G. (2000). Biofibres, biodegradable polymers and biocomposites: An overview, *Macromol. Mater. Eng.,* **276–277**, pp. 1–24.
13. Bordes, P., Pollet, E., and Averous, L. (2009). Nano-biocomposites: Biodegradable polyester/nanoclay systems, *Prog. Polymer Sci.*, **34**, 125–155.
14. Park, H., Shalaby. W. S. W. (eds.) (1993). Biodegradable Hydrogels for Drug Delivery (CRC Press, Boca Raton), pp. 107–108.

15. Tănase, E. E., Râpă, M., Popa, O. (2014). Biopolymers based on renewable resources-A review, *Sci. Bull. Series F. Biotechnol.*, **18**, pp. 188–195.
16. Bakiyalakshmi, S. V., Valli, V., and Delinta Linu Swarnila, R. (2016). Isolation and application of chitin and chitosan from crab shell, *Int. J. Curr. Microbiol. Appl. Sci.*, **3**, pp. 91–99.
17. Duo-Chuan, L. (2006). Review of fungal chitinases, *Mycopathologia*, **161**, pp. 345–360.
18. Nopakarn, P., Abhinaya, P., Yano, S., Marunu, W., and Takashi, T. (2002). Utilization of shrimp fish shell waste was a substrate for solid state cultivation of *Aspergillus sp S1-13*, *J. Biosci. Bioeng.*, **93**, pp. 550–556.
19. Chang, W., Chen, Y., and Jao, C. (2007). Antifungal activity and enhancement of plant growth by Bacillus cereus grown on shellfish chitin wastes, *Biores. Technol.*, **98**, pp. 1224–1230.
20. Synowiecki, J., and Al-Khateeb, N. A. A. Q. (2000). The recovery of protein hydrolysate during enzymatic isolation of chitin from shrimp crangon processing discards, *Food Chem.*, **68**, pp. 147–152.
21. Gopalan, N. K., and Dufresne, A. (2003). Crab shell chitin whisker reinforced natural rubber nanocomposites. 1. Processing and swelling behavior, *Biomacromolecules*, **4**, pp. 657–665.
22. Dutta, P. K., dutta, J., and Tripathi, V. S. (2004). Chitin and chitosan: Chemistry, properties and applications, *J. Sci. Ind. Res.*, **63**, pp. 20–31.
23. Bueter, C. L., Specht, C., and Levitz, S. (2013). Innate sensing of chitin and chitosan, *PLoS Pathog.*, **9**, p. e1003080.
24. Pillai, C. K. S., Paul, W., and Sharma, C. P. (2009). Chitin and chitosan polymers Chemistry, solubility and fiber formation, *Prog. Polym. Sci.*, **34**, pp. 641–678.
25. Muzzarelli, R. A. A. (1977). *Chitin*, Pergamon Press, Oxford, UK.
26. Sahoo, S., Sahoo, R., and Nayak, P. L. (2013). Synthesis and characterization of gelatinchitosan nanocomposite to explore the possible use as drug delivery vehicle, *Eur. Sci. J.*, **9**, p. 135.
27. Peter, M. G. (2002). Chitin and chitosan from animal sources. In: *Biopolymers*, (Polysaccharides II), De Baets, S., Vandamme, E. J., and Steinbuchel, A. (eds.), Wiley-VCH, Weinheim, 2, Chapter 15.
28. Gartman, O., and Vorobyeva, V. (2013). Technology and properties of chitosan from the gammarus, *Fundam. Res.*, **6**, pp. 1188–1192.

29. Prashanth, K. V. H., and Tharanathan, R. N. (2007). Chitin/chitosan modifications and their unlimited application potential—an overview, *Trends Food Sci. Technol.*, **18**, pp. 117–31.

30. Kurita, K. (1986). Chemical modifications of chitin and chitosan. In *Chitin in Nature and Technology*, Muzzarelli, R., Jeuniaux, C., Gooday, G. W. (ed.), pp. 287–293.

31. Jayakumar, R., Prabaharan, M., Nair, S., Tokura, S., Tamura, H., and Selvamurugan, N. (2010). Novel carboxymethyl derivatives of chitin and chitosan materials and their biomedical applications, *Prog. Mater. Sci.*, **55**, pp. 675–709.

32. Jeong, Y., Kim, D., Jang, M., and Nah, J. (2008). Preparation and spectroscopic characterization of methoxy poly(ethylene glycol)-grafted water-soluble chitosan, *Carbohydr. Res.*, **343**, pp. 282–289.

33. Jalal Zohuriaan-Mehr, M. (2005). Advances in chitin and chitosan modification through graft copolymerization: A comprehensive review, *Iran. Polym. J.*, **14**, pp. 235–265.

34. Arvanitoyannis, I., Nakayama, A., and Aiba, S. (1998). Chitosan and gelatin based edible films: State diagrams, mechanical and permeation properties, *Carbohydr. Polym.*, **37**, pp. 371–382.

35. Ioannis Arvanitoyannis, I. S., Kolokuris, I., Atsuyoshi Nakayama, A., Noboru Yamamoto, N., and Aiba, S. (1997). Physico-chemical studies of chitosan-poly(viny1alcohol) blends plasticized with sorbitol and Sucrose, *Carbohydr. Polym.*, **34**, pp. 9–19.

36. Inmaculada, A., Ruth, H., and Angeles, H. (2010). Chitosan amphiphilic derivatives chemistry and applications, *Org. Curr. Chem.*, **14**, pp. 308–333.

37. Hui, L., Yumin, D., Xiaohui, W., and Liping, S. (2004). Chitosan kills bacteria through cell membrane damage, *Int. J. Food Microbiol.*, **95**, pp. 147–155.

38. Tamara, K., Michelle, W., Yasmin, A., and Shelley, D. M. (2008). Improving the microenvironment for enzyme immobilization at electrodes by hydrophobically modifying chitosan and Nafion polymers, *J. Membr. Sci.*, **311**, pp. 81–88.

39. Elena, R., Lupei, M., Ciolacu, F. L., and Desbrières, J. (2011). Synthesis and characterization of n-alkyl chitosan for papermaking applications, *Cellulose Chem. Technol.*, **45**, pp. 619–625.

40. Muzzarelli, R. A. (1988). Carboxymethylated chitins and chitosan, *Carbohydr. Polym.*, **8**, pp. 1–21.

41. Thanou, M., Nihot, M. T., Jansen, M., Verhoef, J. C., and Junginger, H. E. (2001). Mono-N-carboxymethyl chitosan (MCC), a polyampholytic chitosan derivative, enhances the intestinal absorption of low molecular weight heparin across intestinal epithelia in vitro and in vivo, *J. Pharm. Sci.,* **90**, pp. 38–46.

42. Chen, Y., and Tan, H. (2006). Crosslinked carboxymethylchitosan-g-poly(acrylic acid) copolymer as a novel superabsorbent polymer, *Carbohydr. Res.,* **341**, pp. 887–896.

43. Kumar, G., Smith, P. J., and Payne, G. F. (1999). Enzymatic grafting of a natural product onto chitosan to confer water solubility under basic conditions, *Biotechnol. Bioeng.,* **63**, pp. 154–165.

44. Yoo, S. H., Lee, J. S., Park, S. Y., Kim, Y. S., Chang, P. S., and Lee, H. G. (2005). Effects of selective oxidation of chitosan on physical and biological properties, *Int. J. Biol. Macromol.,* **35**, pp. 27–31.

45. Islam, M. M., Masum, S. M., Rahman, M. M., Mollab, M. A. I., Shaikh, A. A., and Roya, S. K. (2011). Preparation of chitosan from shrimp shell and investigation of its properties, *Int. J. Basic Appl. Sci.,* **11**, pp. 77–80.

46. Hasri, H. (2010). Prospects modified chitosan and chitosan as a promising natural biopolymers, *J. Chem.,* **11**, pp. 1–10.

47. Khoushab, F., and Yamabhai, M. (2010). Chitin research revisited, *Mar. Drugs,* **8**, pp. 1988–2012.

48. Aranaz, I., Mengibar, M., Harris, R., Panos, I., Miralles, B., Acosta, N., Galed, G., and Heras, A. (2009). Functional characterization of chitin and chitosan, *Curr. Chem. Biol.,* **3**, pp. 203–230.

49. Olteanu, C. E., and Enescu, D. (2007). Chitosan involved tissue engineering and regenerative medicine, *Roum Biotechnol Lett.,* **12**, pp. 3217–3233.

50. Zhu, X., Chian, K. S., Chan-Park, M. B. E., and Lee, S. T. (2005). Effect of argon-plasma treatment on proliferation of human-skin-derived fibroblast on chitosan membrane in vitro, *J. Biomed. Mater. Res. Part A,* **73A**, pp. 264–274.

51. Francesko, A., and Tzanov, T. (2011). Chitin, chitosan and derivatives for wound healing and tissue engineering, *Adv. Biochem. Eng. Biotechnol.,***125**, pp. 1–27.

52. Dutta, P., Rinki, K., and Dutta, J. (2011). Chitosan: A promising biomaterial for tissue engineering scaffolds. Chitosan for biomaterials II. In: *Advances in Polymer Science,* Jayakumar, R.,

Prabaharan, M., and Muzzarelli, R. A. A. (eds.), Heidelberg: Springer Berlin, pp. 45–79.

53. Jagur-Grodzinski, J. (2003). Biomedical applications of polymers 2001–2002, *e-Polym.*, Paper No. 12.
54. Madhavan, P. (1992). Chitin, chitosan and their novels applications. Science lecture series, CIFT, Kochi, p. 1.
55. Pal, J., Verma, H. O., Munka, V. K., Maurya, S. K., Roy, D., and Kumar, J. (2014). Biological method of chitin extraction from shrimp waste an eco-friendly low cost technology and its advanced application, *Int. J. Fisheries Aquatic Studies*, **6**, pp. 104–107.
56. Rathke, T. (1993). Determination of the degree of N-deacteylation in chitin and chitosan as well as their monomer sugar ratios by near infrared spectroscopy, *J. Polym. Sci. Polym. Hem.*, **31**, pp. 749–753.
57. Kurita, K. (2001). Controlled functionalization of polysaccharide chitin, *Prog. Poly Sci.*, **26**, pp. 1921–1971.
58. Varsha Panchal., Vyas, B., Chauhan, C. S., Goyal, P. K., and Sarangdevot, Y. S. (2015). Chitosan as a natural polymer: An overview, *Int. J. Pharm. Erudition.*, **5**, pp. 1–8.
59. Lee, V. F. (1974). Solution and shear properties of chitin and chitosan. Ph.D. Dissertation, university of Washington, university microfilms. Ann Arbor, MI, USA, Microfilm, **446**: 74–29.
60. No, H. K., Meyers, S. P., Prinyawiwatkul, W., and Xu, Z. (2007). Application of chitosan for improvement of quality and shelf life of foods-A review, *J. Food Sci.*, **72**, pp. 87–100.
61. Cho, Y. I., No, H. K., and Meyers, S. P. (1998). Physicochemical characteristics and functional properties of various commercial chitin and chitosan products, *J. Agric. Food Chem.*, **46**, pp. 3839–3843.
62. Fai, A., Stamford, T., Stamford-Arnaud, T., Santa-Cruz, P., Silva, M., Campos-Takaki, G., and Stamford, T. (2011). Physico-chemical characteristics and functional properties of chitin and chitosan produced by Mucor circinelloides using Yam Bean as substrate, *Molecules*, **16**, pp. 7143–7154.
63. Dash, M., Chiellini, F., Ottenbrite, R., and Chiellini, E. (2011). Chitosan—A versatile semi-synthetic polymer in biomedical applications, *Prog. Polym. Sci.*, **36**, pp. 981–1014.
64. Amidi, M., Mastrobattista, E., Jiskoot, W., and Hennink, W. (2010). Chitosan-based delivery systems for protein therapeutics and antigen, *Adv. Drug Deliv. Rev.*, **62**, pp. 59–82.

65. Agarwal, P., Strijkers, G. J., and Nicolay, K. (2010). Chitosan-based systems for molecular imaging, *Adv. Drug Deliv. Rev.*, **62**, pp. 42–58.

66. Kumirska, J., Weinhold, M. X., Thoming, J., and Stepnowski, P. (2011). Biomedical activity of chitin/chitosan based materials influence of physicochemical properties apart from molecular weight and degree of *N*-acetylation, *Polymers*, **3**, pp. 1875–1901.

67. Aranaz, I., Mengibar, M.., Harris, R., Panos, I., Miralles, B., Acosta, N., Galed, G., and Heras, Á. (2009). Functional characterization of chitin and chitosan, *Curr. Chem. Biol.*, **3**, pp. 203–230.

68. Weinhold, M. X., Sauvageau, J. C. M., Keddig, N., Matzke, M., Tartsch, B., Grunwald, I., Kubel, C., Jastorff, B., and Thoming, J. (2009). Strategy to improve the characterization of chitosan for sustainable biomedical applications: SAR guided multi-dimensional analysis, *Green Chem.*, **11**, pp. 498–509.

69. Huang, M., Khor, E., and Lim, L. (2004). Uptake and cytotoxicity of chitosan molecules and nanoparticles: Effects of molecular weight and degree of deacetylation, *Pharm. Res.*, **21**, pp. 344–353.

70. Tomihata, K., and Ikada, Y. (1997). In vitro and in vivo degradation of films of chitins and its deacetylated derivatives, *Biomaterials*, **18**, pp. 567–575.

71. Zhang, H., and Neau, S. H. (2001). In vitro degradation of chitosan by a commercial enzyme preparation: Effect of molecular weight and degree of deacetylation, *Biomaterials*, **22**, pp. 1653–1658.

72. Aiba, S. (1992). Studies on chitosan: Lysozymic hydrolysis of partially N-acetylated chitosan, *Int. J. Biol. Macromol.*, **14**, pp. 225–228.

73. Sashiwa, H., Saimoto, H., Shigemasa, Y., Ogawa, R., and Tokura, S. (1991). Distribution of the acetamide group in partially deacetylated chitins, *Carbohydr. Polym.*, **16**, pp. 291–296.

74. Shigemasa, Y., Saito, K., Sashiwa, H., and Saimoto, H. (1994). Enzymatic degradation of chitins and partially deacetylated chitins, *J. Biol. Macromol.*, **16**, pp. 43–49.

75. Knapczk, J., Krowczynski, L., Pawlik, B., and Liber, Z. (1984). Pharmaceutical dosage form in chitosan. In: *Chitin and Chitosan: Sources, Chemistry, Biochemistry, Physical Properties and Applications*, Skjak-Braek, G., Anthonsen, T., and Sanford, P., (eds.), London: Elsevier Applied Science, pp. 665–669.

76. Chatelet, C. (2001). Influence of the degree of acetylation on some biological properties of chitosan films, *Biomaterials*, **2**, pp. 261–268.

77. Jian, Y., Feng, T., Zheng, W., Qing, W., Yan-Jun, Z., and Shi-Qian, C. (2008). Effect of chitosan molecular weight and deacetylation degree on hemostasis, *J. Biomed. Mater. Res. B*, **84B**, pp. 131–137.

78. Okamoto, Y., Kawakami, K., Miyakate, K., Morimoto, M., Shigesama, Y., and Minami, S. (2002). Analgesic effects of chitin and chitosan, *Carbohydr. Polym.*, **49**, pp. 249–252.

79. Tsukada, K., Matsumoto, T., Aizawa, K., et al. (1990). Antimetastatic and growth-inhibitory effects of N-acetylchitohexaose in mice bearing Lewis lung carcinoma. *Jpn. J. Cancer Res.*, **81**, pp. 259–265.

80. Kulikov, S., Chirkov, S., Il'ina, A., Lopatin, S., and Varlamov, V. (2006). Effect of the molecular weight of chitosan on its antiviral activity in plants, *Appl. Biochem. Microbiol.*, **42**(2), pp. 200–203.

81. Lopez-Carballo, G., Gomez-Estaca, J., Catala, R., Hernandez-Munoz, P., and Gavara, R. (2012). Active antimicrobial food and beverage packaging, *Emerg. Food Packaging Technol.*, **3**, pp. 27–54.

82. Li, X., Li, X., Ke, B., Shi, X., and Du, Y. (2011). Cooperative performance of chitin whisker and rectorite fillers on chitosan films, *Carbohydr. Polym.*, **85**, pp. 747–752.

83. Sudharshan, N. R., Hoover, D. G., and Knorr, D. (1992). Antibacterial action of chitosan, *Food Biotechnol.*, **6**, pp. 257–272.

84. Xiao, B., Wan, Y., Zhao, M., Liu, Y., and Zhang, S. (2011). Preparation and characterization of antimicrobial chitosan-N-arginine with different degrees of substitution, *Carbohydr. Polym.*, **83**, pp. 144–150.

85. Leceta, I., Guerrero, P., Ibarburu, I., Duenas, M. T., and de la Caba, K. (2013). Characterization and antimicrobial analysis of chitosan-based films, *J. Food Eng.*, **116**, pp. 889–899.

86. Tran, M. Q. (2014). Radiation processing of bio-polymers for agriculture applications, In: *Radiation Processed Materials in Products from Polymers for Agricultural Applications* (book chapter), International Atomic Energy Agency, IAEA-TECDOC-1745, Vienna.

87. Saima, M. K., and Roohi, I. Z. A. (2013). Isolation of novel chitinolytic bacteria and production optimization of extracellular chitinase, *J. Genetic Eng. Biotechnol.*, **11**, pp. 39–46.

88. Chandler, D., Bailey, A., Tatchell, G., Davidson, G., Greaves, J., and Grant, W. (2011). The development, regulation and use of biopesticides for integrated pest management, *Philos. Trans. R. Soc.*, **366**, pp. 1987–1998.

89. Lowe, A., Rafferty-McArdle, S. M., and Cassells, A. C. (2012). Effects of AMF- and PGPR-root inoculation and a foliar chitosan spray in

single and combined treatments on powdery mildew disease in strawberry, *Agric. Food Sci.,* **21**, pp. 28–38.

90. Sharp, R. G. (2013). A review of the applications of chitin and its derivatives in agriculture to modify plant-microbial interactions and improve crop yields, *Agronomy,* **3**, pp. 757–793.
91. Sun, B., Zhang, L., Yang, L., Zhang, F., Norse, D., and Zhu, Z. (2012). Agricultural non-point source pollution in China: Causes and mitigation measures, *Ambio,* **41**, pp. 370–379.
92. Katiyar, D., Hemantaranjan, Singh, B., Bhanu, N. (2014). Future perspective in crop protection chitosan and its oligosaccharides, *Adv. Plants Agric. Res.,* **1**, pp. 1–8.
93. Bautista-Baños, S., Hernández-Lauzardo, A., Velázquez-del Valle, M., Hernández-López, M., Ait Barka, E., Bosquez-Molina, E., and Wilson, C. (2006). Chitosan as a potential natural compound to control pre and postharvest diseases of horticultural commodities, *Crop. Prot.,* **25**, pp. 108–118.
94. Cousin, M. (1996). Chitin as a measure of mold contamination of agricultural commodities and foods, *J. Food Protection.,* **59**, pp. 73–81.
95. Donald, W. W., and Mirocha, C. J. (1977). Chitin as a measure of fungal growth in stored corn and soybean seed, *In. Cereal Chem.,* **54**, pp. 466–474.
96. Bishop, R., Duncan, C., Evancho, G., and Young, H. (1982). Estimation of fungal contamination in tomato products by a chemical assay for chitin, *J. Food Sci.,* **47**, pp. 437–439.
97. Ma, P., and He, L. Q. (2001). Progress of chitosan in suppression of plant diseases, *Natl. Prod. Res. Development.,* **13**, pp. 82–86.
98. Bautista-Banosa, S., Hernandez L Opez, M., Bosquez-Molina, E., and Wilson, C. L. (2003). Effects of chitosan and plant extracts on growth of Colletotrichum gloeosporioides, anthracnose levels and quality of papaya fruit, *Crop Protection.,* **22**, pp. 1087–1092.
99. Kittur, F. S., Saroja, N., Habibunnisa and Tharanathan, R. N. (2001). Polysaccharide-based composite coating formulations for shelf-life extension of fresh banana and mango, *Eur. Food Res. Technol.,* **213**, pp. 306–311.
100. Srinivasa, P. C., Baskaran, R., Ramesh, M. N., Prashanth, K. V. H., and Tharanathan, R. N. (2002). Storage studies of mango packed using biodegradable chitosan film, *Eur. Food Res. Technol.,* **215**, pp. 504–508.

101. Li., H., and Yu, Y. (2001). Effect of chitosan on incidence of brown rot quality and physiological attributes of postharvest peach fruit, *J. Sci. Food. Agric.*, **81**, pp. 269–274.

102. Meighani, H., Ghasemnezhad, M., and Bakhshi, D. (2015). Effect of different coatings on post-harvest quality and bioactive compounds of pomegranate (*Punica granatum* L.) fruits, *J. Food Sci. Technol.*, **52**, pp. 4507–4514.

103. Abdel-Mawgoud, A. M. R., Tantawy, A. S., El-Nemr, M. A., and Sassine, Y. N. (2010). Growth and yield responses of strawberry plants to chitosan application, *Eur. J. Sci. Res.*, **39**, pp. 161–168.

104. Romanazzi, G., Karabulut, O. A., and Smilanick, J. L. (2007). Combination of chitosan and ethanol to control postharvest gray mold of table grapes, *Postharvest Biol. Technol.*, **45**, pp. 134–140.

105. Wang, S. Y., and Gao, H. (2012). Effect of chitosan based edible coating on antioxidants, antioxidant enzyme system, and postharvest fruit quality of strawberries, *Food Sci. Tech.*, **52**, pp. 71–19.

106. Li, J., Yan, J., Wang, J., Zhao, Y., Cao, J., and Jiang, W. (2010). Effects of chitosan coating in oxidative stress in bruised Yali Pears, *Int. J. Food Sci. Tech.*, **45**, pp. 2149–2154.

107. Lizárraga-Paulín, E. G., Miranda-Castro, S. P., Moreno-Martínez, E., Lara-Sagahón, A.-V., and Torres-Pacheco, I. (2013). Maize seed coatings and seedling sprayings with chitosan and hydrogen peroxide: Their influence on some phenological and biochemical behaviors, *J. Zhejiang Univ. Sci. B.*, **14**, pp. 87–96.

108. Zhou, Y. G., Yang, Y. D., Qi, Y. G., Zhang, Z. M., Wang, X. J., and Hu, X. J. (2002). Effects of chitosan on some physiological activity in germinating seed of peanut, *J. Peanut Sci.*, **31**, pp. 22–25.

109. Ziani, K., Ursúa, B., and Maté, J. I. (2010). Application of bioactive coatings based on chitosan for artichoke seed protection, *Crop Prot.*, **29**, pp. 853–859.

110. Hirano, S., Yamamoto, T., Hayashi, M., Nishida, T., and Inui, H. (1990). Chitinase activity in seeds coated with chitosan derivatives, *Agric. Biol. Chem.*, **54**, pp. 2719–2720.

111. Cline, S. D., and Coscia, C. J. (1988). Stimulation of sanguinarine production by combined fungal elicitation and hormonal deprivation in cell suspension cultures of Papaver bracteatum, *Plant Physiol.*, **86**, pp. 161–165.

112. Darvill, A. G., and Albersheim, P. (1984). Phytoalexins and their elicitors–A defense against microbial infection in plants, *Ann. Rev. Plant Physiol.*, **35**, pp. 243–275.

113. Funk, C., and Brodelius, P. (1990). Influence of growth regulators and an elicitor on phenylpropanoid metabolism in suspension cultures of Vanilla planifolia, *Phytochemistry*, **29**, pp. 845–848.

114. Bhattacharya, D., Nagpure, A., and Gupta, R. (2007). Bacterial chitinases: Properties and potential, *Crit. Rev. Biotechnol.*, **27**, pp. 21–28.

115. Doares, S., Syrovets, T., Weiler, E., and Ryan, C. (1995). Oligogalacturonides and chitosan activate plant defensive genes through the octadecanoid pathway, *PNAS*, **92**, pp. 4095–4098.

116. Eikemo, H., Stensvand, A., and Tronsmo, A. (2003). Induced resistance as a possible means to control diseases of strawberry caused by *Phytophthora spp.*, *Plant Disease*, **87**, pp. 345–350.

117. Bell, A., Hubbard, J., Liu, L., Davis, R., and Subbarao, K. (1998). Effects of chitin and chitosan on the incidence and severity of fusarium yellows of celery, *Plant Disease*, **82**, pp. 322–328.

118. Kunkel, B. N., and Brooks, D. M. (2002). Cross talk between signaling pathways in pathogen defense, *Curr. Opin. Plant Biol.*, **5**, pp. 325–331.

119. Grennan, A. K. (2006). Plant response to bacterial pathogens. Overlap between innate and gene-for-gene defense response, *Plant Physiol.*, **142**, pp. 809–811.

120. Dangl, J. L., and Jones, J. D. G. (2001). Plant pathogens and integrated defence responses to infection, *Nature*, **411**, pp. 826–833.

121. Shibuya, N., and Minami, E. (2001). Oilgosaccharide signaling for defence responses in plant, *Physiol. Mol. Plant Pathol.,* **59**, pp. 223–233.

122. Weller, D. M., Raaijmakers, J. M., McSpaddenGardener, B. B., and Thomashow, L. S. (2002). Microbial populations responsible for specific soil suppressiveness to plant pathogens, *Annu. Rev. Phytopathol.*, **40**, pp. 309–348.

123. Sarathchandra, S. U., Watson, R. N., Cox, N. R., diMenna, M. E., Brown, J. A., Burch, G., and Neville, F. J. (1996). Effects of chitin amendment of soil on microorganisms, nematodes, and growth of white clover (*Trifolium repens* L.) and perennial ryegrass (*Lolium perenne* L.), *Biol. Fertil. Soils.,* **22**, pp. 221–226.

124. Green, S. J., Inbar, E., Michel, F. C Jr., Hadar, Y., and Minz, D. (2006). Succession of bacterial communities during early plant development: Transition from seed to root and effect of compost amendment, *Appl. Environ. Microbiol.*, **72**, pp. 3975–3983.

125. Radwan, M. A., Farrag, S. A. A., AbuElamayem, M. M., and Ahmed, N. S. (2012). Extraction, characterization, and nematicidal activity of

chitin and chitosan derived from shrimp shell waste, *Biol. Fertil. Soils.*, **48**, pp. 463–468.

126. Mendes, R., Kruijt, M., Debruijn, I., Dekkers, E., Vandervoort, M., Schneider, J. H. M., Piceno, Y. M., DeSantis, T. Z., Andersen, G. L., Bakker, P. A. H. M., and Raaijmakers, J. M. (2011). Deciphering the rhizosphere microbiome for disease-suppressive bacteria, *Science*, **332**, pp. 1097–1100.

127. Chen, H., Seguin, P., Archambault, A., Constan, L., and Jabaji, S. (2009). Gene expression and isoflavone concentrations in soybean sprouts treated with chitosan, *Crop Sci.*, **49**, p. 224.

128. Lin, W., Hu, X., Zhang, W., Rogers, W. J., and Cai, W. (2005). Hydrogen peroxide mediates defense responses induced by chitosan of different molecular weights in rice, *J. Plant Physio.*, **162**, pp. 937–944.

129. Li, Y., Yin, H., Wang, Q., Zhao, X. M., Du, Y. G., and Li, F. L. (2009). Oligochitosan induced *Brassica napus* L. production of NO and H_2O_2 and their physiological function, *Carbohyd. Polym.*, **75**, pp. 612–617.

130. Dzung, N., Khanh, V., and Dzung, T. (2011). Research on impact of chitosan oligomers on biophysical characteristics, growth, development and drought resistance of coffee, *Carbohyd. Polym.*, **84**, pp. 751–755.

131. Zhao, X. M., She, X. P., Yu, W., Liang, X. M., and Du, Y. G. (2007). Effects of oligochitosan on tobacco cells and role of 8 endogenous nitric oxide burst in the resistance of tobacco to TMV, *J. Plant Pathol.*, **89**, pp. 55–65.

132. Zhang, H., Zhao, X., Yang, J., Yin, H., Wang, W., Lu, H., and Du, Y. (2011). Nitric oxide production and its functional link with OIPK in tobacco defense response elicited by chitooligosaccharide, *Plant Cell Rep.*, **30**, pp. 1153–1162.

133. Kim, S., and Rajapakse, N. (2005). Enzymatic production and biological activities of chitosan oligosaccharides (COS): A review, *Carbohyd. Poly.*, **62**, pp. 357–368.

134. Yin, H., Li, Y., Zhang, H. Y., Wang, W. X., Lu, H., Grevsen, K., Zhao, X., and Du, Y. (2013). Chitosan oligosaccharides triggered innate immunity contributes to oilseed rape resistance against *Sclerotinia Sclerotiorum*, *Int. J. Plant Sci.*, **174**, pp. 722–732.

135. Khan, W. M., Prithiviraj, B., and Smiyh, D. L. (2002). Effect offoliar application of chitin oligosaccharides on photosynthesis of maize and soybean, *Photosynthetica.*, **40**, pp. 621–624.

136. Chibu, H., and Shibayama, H. (2003). Effects of chitosan application on the growth of several crops. In: *Chitin and Chitosan in Life Science*,

In: Uragami, T., Kurita, K., and Fukamizo, T. (eds.), Yamaguchi, Japan, pp. 235–239.

137. Gornik, K., Grzesik, M., and Duda, B. R. (2008). The effect of chitosan on rooting of gravevine cuttings and on subsequent plant growth under drought and temperature stress, *J. Fruit Ornamental Plant Res.*, **16**, pp. 333–343.

138. Hadrami, A. E., Adam, L. R., Hadrami, I. E., and Daayf, F. (2010). Chitosan in plant protection, *Mar. Drugs*, **8**, pp. 968–987.

139. Cretoiu, M. S., Korthals, G. W., Visser, J. H. M., and van Elsasa, V. J. D. (2013). Chitin amendment increases soil suppressiveness toward plant pathogens and modulates the actinobacterial and oxalobacteraceal communities in an experimental agricultural field, *Appl. Environ. Microbiol.*, **79**, pp. 5291–5301.

140. El Ghaouth, A., Arul, J., Asselin, A., and Benhamou, N. (1992). Antifungal activity of chitosan on post-harvest pathogens: Induction of morphological and cytological alterations in *Rhizopus stolonifer*, *Mycol. Res.*, **96**, pp. 769–779.

141. El-Ghaouth, A., Smilanick, J., and Wilson, C. (2000). Enhancement of the performance of Candida saitoana by the addition of glycolchitosan for the control of postharvest decay of apple and citrus fruit, *Postharvest Biol. Technol.*, **19**, pp. 103–110.

142. Mondal, M. M. A., Malek, M. A., Puteh, A. B., Ismail, M. R., Ashrafuzzaman, M., and Naher, L. (2012). Effect of foliar application of chitosan on growth and yield in okra, *Aust. J. Crop Sci.*, **6**, pp. 918–921.

143. Sivakumar, D., Sultanbawa, Y., Ranasingh, N., and Wijesundera, R. L. C. (2005). Effect of the combined application of chitosan and carbonate salts on the incidence of anthracnose and on the quality of papaya during storage, *J. Hortic. Sci. Biotechnol.*, **80**, pp. 447–452.

144. Boonlertnirun, S., Boonraung, C., and Suvanasara, R. (2008). Application of chitosan in rice production, *J. Metals Mater. Minerals*, **18**, pp. 47–52.

145. Utsunomiya, N., and Kinai, H. (1994). Effect of chitosan-oligosaccharides soil conditioner on the growth of passionfruit, *J. Jpn. Soc. Hortic. Sci.*, **64**, pp. 176–177.

146. Austin, P., Brine, C., Castle, J., and Zikakis, J. (1981). Chitin: New facets of research. *Science.*, **212**, pp. 749–753.

147. Knorr, D. (1986). Nutritional quality, food processing, and biotechnology aspects of chitin and chitosan: A review, *Proc. Biochem.*, **6**, pp. 90–92.

148. Chibu, H., Shibayama, H., and Arima, S. (2002). Effects of chitosan application on the shoot growth of rice and soybean, *Jap. J. Crop. Sci.*, **71**, pp. 206–211.

149. Boonlertnirun, S., Boonlertnirun, K., and Sooksathan, I. (2005). Effect of chitosan application on growth and yield of rice (Oryza sativa) var. Suphunburi 1, *Proceedings of 43rd Kasetsart University Annual Conference*, Thailand, **1**, pp. 37–43.

150. Rehim, H. A. A., Hegazy, E. A., and El-Barbary, A. M. (2009). Radiation modification of natural polysaccharides for applacation in agriculture, *Polymer,* **50**, pp. 1952–1957.

151. Zeng, D., Luo, X., and Tu, R. (2012). Application of bioactive coatings based on chitosan for soybean seed protection, *Int. J. Carbohydr. Chem.*, **2012**, pp. 1–5.

152. Ait Barka, E., Eullaffroy, P., Clément, C., and Vernet, G. (2003). Chitosan improves development, and protects *Vitis vinifera L.* against *Botrytis cinerea*, *Plant Cell Rep.*, **22**, pp. 608–614.

153. Pornpeanpakdee, P., Pichyangkura, R., Chadchawan, S., and Limpanavech, P. (2005). Chitosan effects on Dendrobium 'Eiskul' Protocorm-like body production. In *Proceedings of the 31st Congress on Science and Technology of Thailand*, Nakornrachaseema, Thailand, 18–20 October 2005, pp. 1–3.

154. Nahar, S. J., Kazuhiko, S., and Haque, S. M. (2012). Effect of polysaccharides including elicitors on organogenesis in protocorm-like body (PLB) of *Cymbidium insigne in vitro, J. Agric. Sci. Technol.,* **2**, pp. 1029–1033.

155. Sauerwein, M., Flores, H. M., Yamazaki, T., and Shimomura, K. (1991). *Lippia dulcis* shoot cultures as a source of the sweet sesquiterpene hernandulcin, *Plant Cell Rep.*, **9**, pp. 663–666.

156. Darwis, D., Puspitasari, T., Iramani, D., Sri Susilowati., D. S., and Pangerteni, D. S. (2014). Preparation of low molecular weight chitosan by radiation and its application for plant growth promoter. In: *Radiation Processed Materials in Products from Polymers for Agricultural Applications* (book chapter) IAEA-TECDOC-1745, Published by International Atomic Energy Agency Vienna International Centre, Vienna, p. 205.

157. Wu, L., Liu, M., and Liang, R. (2008). Preparation and properties of a double-coated slow-release NPK compound fertilizer with superabsorbent and water-retention, *Bioresour Technol.,* **99**, pp. 547–554.

158. Corradini, E., de Moura, M., and Mattoso, L. (2010). A preliminary study of the incorparation of NPK fertilizer into chitosan nanoparticles, *Express Polym. Lett.*, **4**, pp. 509–515.

159. Hamid, N. N. A., Mohamad, N., Hing, L. Y., Dimin, M. F., Azam, M. A., Hassan, M. H. C., Ahmad, M. K. S. M., and Shaaban, A. (2013). The effect of chitosan content to physical and degradation properties of biodegradable urea fertilizer, *J. Sci. Innovative Res.*, **2**, pp. 893–902.

160. dos Santos, B. R., Bacalhau, F. B., Pereira Tdos, S., Souza, C. F., and Faez, R. (2015). Chitosan-montmorillonite microspheres: A sustainable fertilizer delivery system, *Carbohydr. Polym.*, **127**, pp. 340–346.

161. Narayanan, A., and Dhamodharan, R. (2015). Super water-absorbing new material from chitosan, EDTA and urea, *Carbohydr. Polym.*, **134**, pp. 337–343.

162. Roshanravan, B., Soltani, S. M., Rashid, S. A., Mahdavi, F., and Yusop, M. K. (2015). Enhancement of nitrogen release properties of urea–kaolinite fertilizer with chitosan binder, *Chem. Speciation Bioavailability*, **27**, pp. 44–51.

163. Cusidó, R. M., Vidal, H., Gallego, A., Abdoli, M., and Palazón, J. (2013). Biotechnological production of taxanes: A molecular approach 75. In: *Recent Advances in Pharmaceutical Sciences,* Muñoz-Torrero, D., Cortés, A., and Mariño, E. L. (eds.), vol. 3, (Transworld Research Network, Trivandrum), pp. 91–107.

164. Gupta, S., Chauhan, D., Mehla, K., Sood, P., and Nair, A. (2010). An overview of nutraceuticals: Current scenario, *J. Basic Clin. Pharm.*, **1**, pp. 55–62.

165. Ramirez-Estrada, K., Vidal-Limon, H., Hidalgo, D., Moyano, E., Golenioswki, M., Cusidó, R. M., and Palazon, J. (2016). Elicitation, an effective strategy for the biotechnological production of bioactive high-added value compounds in plant cell factories, *Molecules*, **21**, p. 182.

166. Orlita, A., Sidwa-Gorycka, M., Paszkiewicz, M., Malinski, E., Kumirska, J., Siedlecka, E. M., Łojkowska, E., and Stepnowski, P. (2008). Application of chitin and chitosan as elicitors of coumarins and furoquinolone alkaloids in *Ruta graveolens L.* (common rue), *Biotechnol. Appl. Biochem.*, **51**, pp. 91–96.

167. Vasyukova, N. I., Zinoveva, S. V., Iiinskaia, L. I., Perekhod, E. A., Chalenko, G. I., et al. (2001). Modulation of plant resistance to diseases by water-soluble chitosan, *Prikladnaia Biokhimiia Mikrobiologiia,* **37**, pp. 115–122.

168. Baque, M. A., Shiragi, M. H. K., Lee, E.-J., and Paek, K.-Y. (2012). Elicitor effect of chitosan and pectin on the biosynthesis of anthraquinones, phenolics and flavonoids in adventitious root suspension cultures of *Morinda citrifolia (L.), Australian J. Crop Sci.*, **6**, pp. 1349–1355.

169. Tao, S., Pang, R., Chen, C., et al. (2012). Synthesis, characterization and slow release properties of O-naphthylacetyl chitosan, *Carbohydr. Polym.*, **88**, pp. 1189–1194.

Chapter 6

Chitosan Applications in Microencapsulation

Berta N. Estevinho and Fernando Rocha

LEPABE, Departamento de Engenharia Química, Faculdade de Engenharia da Universidade do Porto, Rua Dr. Roberto Frias, 4200-465 Porto, Portugal

berta@fe.up.pt

Nowadays, biopolymers such as chitosan have attracted the attention of the scientific community as matrixes for the microencapsulation, immobilization, or controlled release of numerous active compounds and have been applied in the pharmaceutical, food, biomedical, chemical, and waste-treatment industries.

On the other hand, microencapsulation has become a challenging approach to design new materials with biotechnological applications and can be used to increase the effectiveness of many compounds in industry allowing the reformulation of a large number of products, improving and giving them better and new properties.

Handbook of Biopolymers: Advances and Multifaceted Applications
Edited by Shakeel Ahmed, Suvardhan Kanchi, and Gopalakrishnan Kumar

ISBN 978-981-4800-17-4 (Hardcover), 978-0-429-02475-7 (eBook)
www.panstanford.com

In the microencapsulation process, the choice of an encapsulating agent used to prepare the microparticles is very important for the encapsulation efficiency and microparticle stability. Particular attention is given to the selection of the encapsulating agent, considering also the characteristics and limitations of the core material and the desired characteristics in the final product.

This chapter shows the applicability of the microencapsulation of different systems with food, biomedical, and technological interest using chitosan. We focus on this natural polymer, which shows intrinsic properties, such as biocompatibility, non-toxicity, and biodegradability. Other properties of chitosan with special interest for the food industry are related to the anticholesterolemic, hypo-cholesterolemic, antimicrobial, and antioxidant properties. Chitosan has been particularly used as a support in the microencapsulation processes for the controlled release of bioactive compounds. Thus, chitosan, in addition to possessing properties beneficial to health, can be considered a feasible biopolymer to be applied in various encapsulation techniques.

6.1 Introduction

This chapter reviews relevant information about the possibility of producing chitosan microparticles with technological interest.

Microencapsulation has become a challenging approach to design new materials with technological applications. It is one of the most important forms of controlled release of substances and allows the utilization of some that otherwise would be unfeasible [1–3].

This technique is disseminated in several fields of the industry, namely in the food industry (e.g. functional compounds, additives, dyes, flavours), pharmaceutical industry for the release of active compounds or vaccines, cosmetics, textile industry and also agricultural applications [3–10].

The food industry is a good example of the applications of the microencapsulation processes and the cost considerations in the food industry are much more stringent than in, for instance, the pharmaceutical or cosmetic industries. In fact, the microencapsulation technology has been used by the food industry for more than 60 years [11]. The search for new, active and healthy products using active and natural ingredients is promoting the development of new formulations and techniques [12, 13]. The number of food substances that were microencapsulated or submitted to other controlled-release techniques has been increasing, for example, flavours [14–17], dyes [18, 19], stabilizers [20], antioxidants [21–24], enzymes [25–29], probiotics [2, 30] and vitamins [31–34], among others.

The microencapsulation process consists of the formation of small "packaging", called microparticles, microspheres or microcapsules, which consist of structures that have one or more ingredient involved or immobilized by one or more polymers. The microencapsulation has several objectives, including the following: (1) reduce the reactivity of the material that is being encapsulated, (2) protect the encapsulated material against evaporation or loss into other medium, (3) facilitate handling, application and storage of the encapsulated material, (4) promote controlled release and (5) mask unpleasant taste and flavour [35].

The microencapsulation using natural biopolymers can emerge as a solution to improve the stability of numerous active and natural ingredients that are very sensitive to oxygen, light, heat and water [3, 12, 27, 36]. Biopolymers can be used as matrixes for the microencapsulation, immobilization or controlled release of numerous active compounds and have been applied in the pharmaceutical, food, biomedical, chemical, and waste treatment industries. The majority of materials used for microencapsulation are bio-based materials such as fats, waxes, carbohydrates, and animal and plant-derived proteins [37, 38].

Lipids such as glycerides, oils, phospholipids, carotenoids and waxes are used in microencapsulation, allowing a barrier creation for the protection of sensitive ingredients against moisture, plus their transport in aqueous media [37].

Proteins can be extracted from animal-derived products (whey proteins, gelatin, casein) and from vegetables (soy proteins, pea proteins, cereal proteins) and are widely used for the encapsulation of active substances. Proteins present several advantages: biocompatibility, biodegradability, good amphiphilic and functional properties such as water solubility and emulsifying and foaming capacity [37].

Carbohydrate polymers/polysaccharides are the most abundant of the major classes of biomolecules [39]. The major advantages of these biopolymers are their good solubility in water and low viscosity at high concentrations compared, for example, with proteins [37]. Examples of polysaccharides used in microencapsulation are starches, maltodextrin, gum arabic, pectin, alginates and chitosan. On the other hand, carbohydrates can be combined with proteins, being a very relevant combination to improve the emulsifying and filmogenic properties during microencapsulation [37].

Chitosan is a widely used biopolymer. Approximately 100 billion tons/year of chitin are produced on the earth [40], of which chitosan is the main derivative. In the recent years, the potential and common applications of chitin and chitosan and their derivatives were estimated to be more than 200 [41]. The main reasons for the increasing attention to this natural product are certainly its interesting intrinsic properties, such as biocompatibility and biodegradability. Chitosan's other properties with special interest for the food industry are related to the anticholesterolemic, hypo-cholesterolemic, antimicrobial and antioxidant properties [41]. Chitosan has also been widely used in the pharmaceutical and food industries [36]. The use of chitosan microparticles as controlled drug delivery systems for drugs and bioactive compounds like flavours has attracted increasing attention since the beginning of the 1990s [42, 43]. Chitosan can also be used as a support for enzyme immobilization [25, 27, 44–46]. It has been particularly used as support in microencapsulation processes for controlled release of bioactive compounds [47] and vitamins [31]. Estevinho et al. (2013) [36] also microencapsulated peach flavour with water-soluble chitosan with good results. Flavours have a large scope of applications in the food industry. However, some of them are very sensitive to ambient or industrial process conditions. The flavour loss during the storage process of foods is a very common occurrence in the food industry, reaching sometimes values of 60–95% when they are in the free form, because they are very volatile, react with other components and are heat and moisture sensitive [48]. Therefore, microencapsulation using biopolymers, namely chitosan, has attracted the interest of the scientific community and industry, namely for food applications [22, 27, 49, 50].

As shown in Fig. 6.1, an increase can be observed in the number of patents and papers published in the recent years according to the source SCOPUS (www.scopus.com), when the words "microencapsulation + biopolymer" and "microencapsulation + chitosan" in the title or abstract or keywords of the documents were searched. More than 690 publications and 5580 patents were found for the combination "microencapsulation + chitosan" and the tendency is to increase.

In the next sections of this chapter, the principles of microencapsulation, main microencapsulation techniques, design of microparticles prepared with different biopolymers, namely chitosan, importance and stability in the microencapsulation process are discussed.

There are several questions to optimize in terms of microencapsulation techniques. According to the core material and its properties, one or other encapsulating agent can be appropriated and the same considerations need to be done for the selection of the microencapsulation method. Microencapsulation methods can be divided in two main types: chemical processes and mechanical processes. The chemical processes include the methods of coacervation, co-crystallization, molecular inclusion and interfacial polymerization. Mechanical methods include methods such as spray-drying, spray chilling/cooling, extrusion and fluidized bed [50, 51].

After the microencapsulation step, the prepared microparticles need to be characterized in terms of shapes (spheres, irregular particles), structures (porous, compact) and physical structures (vitreous or crystalline) [25, 31, 52].

Finally, some considerations about the controlled release from biopolymer microparticles will be done, namely the discussion of mathematical models. With a properly designed controlled-release system, the compound is released at the desired site and time, and at the desired rate [13, 53].

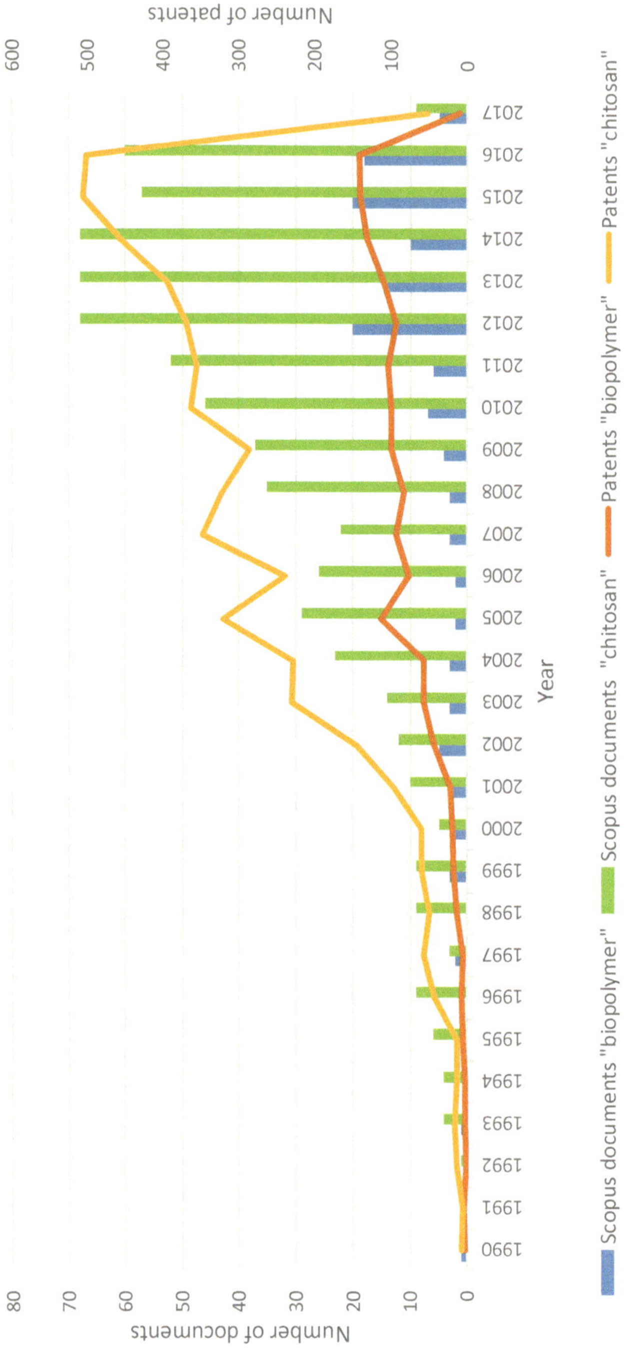

Figure 6.1 Number of patents and papers (documents) published in the last years according to the source SCOPUS (www.scopus.com) containing the words "microencapsulation + biopolymers" and "microencapsulation + chitosan" in the title or abstract or keywords of the document. (The data of 2017 is counted only till 14 February 2017).

6.2 Microencapsulation

The contemporary industrial production of foods, supplements, nutraceuticals, cosmetics, and health products is more sophisticated and involves, for example, the inclusion of bioactive compounds with potential health benefits; however, the stability of bioactive compounds is a critical parameter for their successful incorporation into various systems.

Microencapsulation can be a solution for this, providing a physical barrier between the core compound and other components of the product. In the microencapsulation technique, liquid droplets, solid particles or gas compounds are entrapped in an encapsulating agent/matrix/wall material. Such compounds are totally involved in a coating material or embedded in a homogeneous or heterogeneous matrix, in order to form small capsules with many different and useful properties [3, 54]. Microcapsules normally are small spheres with diameters between a few micrometres and a few millimetres [3, 55].

Microcapsules can be divided into two parts: the core and the shell/wall/encapsulating agent. The compound to be encapsulated (core material) usually stays in the core of the capsule surrounded by the encapsulating agent or dispersed in one matrix containing the encapsulating agent. The encapsulating agent protects the core permanently or temporarily from the external atmosphere [3, 54, 56].

The morphology of microcapsules depends mainly on the core material and the deposition process of the encapsulating agent. Microcapsules may have regular or irregular shapes, and based on their morphology can be classified as mononuclear, polynuclear and matrix types. Some types of microcapsules may have more than one encapsulating agent in its constitution, presenting different walls with different chemical and physical properties. The different types of microparticles can be produced from a wide range of encapsulating agents and by a large number of microencapsulation processes [3, 54, 56].

The selection of the microencapsulation process and the encapsulating agent are interdependent. So, based on the encapsulating agent or process applied, the appropriate process or encapsulating agent is selected. In the next two sections, the different microencapsulation methods and encapsulating agents will be presented and discussed.

6.3 Microencapsulation Methods

In recent years, new encapsulation methods have been developed and the old ones have been improved. There are several encapsulation techniques, with different adaptations and specificities, considering the type of microparticle desired (size and shape), the chemical and physical properties of the compound and of the encapsulating agent, the type of controlled release and the scale of production [13, 54, 57] (Table 6.1). Table 6.1 lists the most important characteristics of four common microencapsulation methods. The two major industrial processes are spray drying and extrusion; also, freeze drying, coacervation and adsorption techniques are intensively used in the industry [58].

The microencapsulation methods can be divided in two main types: chemical processes and mechanical processes. The chemical methods are distinguished by the reactivity of the external wall material (encapsulating agent/shell) surrounding the encapsulated core material. The physical methods are characterized by the use of temperature and pressure to drive formation of the wall that coats the core material [59]. In the next subsections, the chemical and mechanical microencapsulation techniques will be presented in detail.

6.3.1 Chemical Processes

The chemical processes include the methods of coacervation, co-crystallization, molecular inclusion and interfacial polymerization [50, 51].

6.3.1.1 Coacervation

The coacervation process consists of the separation from a solution of colloid particles, which then agglomerate into a separate, liquid phase called coacervate. Generally, the core material used in the coacervation must be compatible with the recipient polymer and be insoluble (or scarcely soluble) in the coacervation medium [58].

Table 6.1 Summary of the most important characteristics of four common microencapsulation methods

		Coacervation	Fluidized bed	Spray drying	Spray cooling
Nature of the core compound	Hydrophilic	Challenging	Straightforward	Unfeasible	Straightforward
	Lipophilic	Straightforward		Straightforward	Unfeasible
	Amphiphilic	Challenging		Challenging	Challenging
	Solid	Straightforward			Straightforward
	Liquid		Unfeasible	Straightforward	Challenging
	> 100 μm		Straightforward	—	Straightforward
	< 100 μm		Unfeasible		
Cost-in-use		Medium	High	Low	Low
Production capacity	Batch wise	0.5 t	1 t	—	—
	Continuous	—	—	2 t/h	5 t/h
Controlled-release mechanism	Thermal	Straightforward	Straightforward	Straightforward	Straightforward
	Time			Challenging	
	Mechanical		Challenging		Challenging
	Digestion		Straightforward		Unfeasible

Source: Adapted from [13].

Coacervation can be simple or complex: Simple coacervation involves only one type of polymer (as encapsulating agent/shell/wall) and the addition of strongly hydrophilic agents to colloidal solution; on the other hand, in complex coacervation, two or more different types of polymers are used [58].

The main advantage of complex coacervation is the production of microcapsules with smaller particle size that ranges from about 1 to 1000 μm. Moreover, compared with other microencapsulation processes complex coacervation gives unusually higher payload of up to 90% for single core and 60% for multicore [60].

However, this method has some problems to solve, namely the optimization of wall material concentration in the emulsification, the evaporation of volatiles, the dissolution of active compound into the processing solvent and product oxidation. Further, complex coacervates are highly unstable and some toxic chemical agents, such as glutaraldehyde, are necessary to stabilize them [12, 58].

6.3.1.2 Molecular inclusion

Molecular inclusion is defined as the result of interactions between compounds in which a smaller guest molecule fits into and is surrounded by the other [58]. It is possible using cyclodextrins [39].

Cyclodextrins are a group of naturally occurring cyclic oligosaccharides with six, seven, or eight glucose residues linked by $\alpha(1 \rightarrow 4)$ glycoside bonds in a cylinder-shaped structure, denominated α-, β- and γ-cyclodextrins, respectively [12]. The external part of the cyclodextrin molecule is hydrophilic, whereas the internal part is hydrophobic. The guest molecules, which are apolar, can be entrapped into the apolar internal cavity through a hydrophobic interaction. This internal cavity of about 0.65 nm diameter permits the inclusion of one or more molecules [61].

6.3.1.3 Co-crystallization

Co-crystallization is another process that has been employed to encapsulate sensitive ingredients [55]. In this method, supersaturated sucrose syrup, utilized as a wall material for the incorporation of bioactive compounds (core material), is maintained at a temperature high enough to avoid crystallization. Core material addition at this stage provides nucleation and triggers the spontaneous crystallization of supersaturated sucrose syrup-core mixture upon cooling. In this way, sucrose crystals can be modified to form aggregates of relatively small, water-soluble crystals that incorporate the core material either within the crystals or by entrapment.

Co-crystallization offers an economical and flexible alternative as the procedure is relatively simple. Numerous products can be encapsulated by co-crystallization process, such as fruit juices, essential oils, and flavours [58]. Although this technique is simple, it does not get much attention in the food industry due to the lack of versatility and sucrose-associated health concerns [55].

6.3.1.4 Interfacial polymerization

The microencapsulation by interfacial polymerization can result from the reaction of two reactive monomers dispersed in one phase, induced to polymerize at the interface or in the dispersed phase and precipitate at the interface [51]. Under these conditions, the surface of the core material is coated by a polymer (encapsulating agent). In this technique, polymers formed by monomers that have preferential solubility for one of the phases are used. Typically, the interfacial polymerization involves the reaction between a diacyl chloride and an amine or alcohol, resulting polyester, polyurea, polyurethane or polycarbonate polymers [51].

6.3.2 Mechanical Processes

The mechanical methods include spray-drying, spray chilling/cooling, extrusion and fluidized bed [50, 51].

6.3.2.1 Spray drying

In spite of all the techniques developed, spray drying is the most common due to its low cost, available equipment and efficiency [62, 63]. It is also a rapid and reproducible technology, allowing easy scale-up, hence justifying the preference in industrial terms [13, 57, 64].

The spray drying technique has been successfully used in industry for several decades, and this process of microencapsulation is one of the oldest encapsulation methods used (since the 1930s) to prepare the first encapsulated flavours [1, 65].

The spray drying technique is flexible, offering substantial variation in matrix microencapsulation, is adaptable to commonly used processing equipment and produces particles of good quality. The microparticles produced are normally matrix type and the mechanisms of release involved are typically controlled by solvents action and by diffusion [66].

The production costs are lower than those associated with most other techniques of microencapsulation. One limitation of the spray drying technology is the type of the encapsulating agent that should be soluble in water at an acceptable level [11, 13]. Spray drying is a unit operation by which a liquid product is atomized in a hot gas (generally air) to instantaneously obtain a powder and generally implies several steps [13].

The first step is dissolving, emulsifying, or dispersing the active compounds (core material) to prepare the "feed solution" that can be a solution, an emulsion or a suspension [12]. Then the following main steps are involved: atomization, formation of the droplet/air contact, evaporation of the water and dry product/humid air separation [57, 67]. Increasing the energy provided to the atomizer decreases the size of the formed droplets. For the same energy amount, the size of formed particles increases with increasing feed rate. On the other hand, the size of particles also increases with the viscosity and surface tension of the feeding liquid [13, 57, 67].

Spray drying produces, depending on the starting feed material and operating conditions, very fine particles (10–50 μm) or large size ones (2–3 mm) [57, 67], which are separated in a cyclone after their formation [57].

According to Madene et al., 2006 [58], some advantages and disadvantages can be presented for the food compounds microencapsulation by a spray drying technique. It is possible to consider as advantages: low operating cost, high quality of capsules in good yield, rapid solubility of the capsules, small size and high stability of the capsules. And as disadvantages: no uniform microcapsules, limitation in the choice of wall material (low viscosity at relatively high concentrations), and very fine powder, which needs additional processing, not good for heat-sensitive material.

6.3.2.2 Spray chilling

Spray chilling is an opposite principle of spray drying; on the other hand, spray chilling and spray cooling are two very similar techniques, being distinguished only by the melting point of the carrier material used. In spray chilling, the most commonly used carriers (encapsulating agents) have a melting point of 32–42°C, while in spray cooling have a melting point of 45–122°C [12]. Commonly spray-chilled carriers for encapsulation include fats, waxes, polyethylene glycols, fatty acids and fatty alcohols [68]. In the spray chilling process, the emulsion is atomized into a spray tower and the shell material is melted and atomized through a pneumatic nozzle into a vessel generally containing a carbon dioxide ice bath (temperature –50°C). Thus, droplets adhere on microparticles and solidify forming a coat film. The process is appropriate for protecting many water-soluble materials that may otherwise be volatilized or damaged during thermal processing [58, 69]. The final characteristics of the microparticles are affected by several processing parameters: size of the orifice of the atomizer, feed flow rate, temperature, compressed air pressure (or wheel speed), air flow rate, and viscosity of the emulsion to be pulverized [12]. Spray chilled products have applications in bakery products, dry soup mixes and foods containing a high level of fat [58].

6.3.2.3 Extrusion

Extrusion is the second major industrial microencapsulation process and was first patented in 1957 [70]. Extrusion is used to encapsulate sensitive compounds and is a relatively low-temperature entrapping method [71]. One of the main advantages of the extrusion method is the stability of the core material against oxidation [70]. The extrusion techniques involve extrusion of a liquid mixture of an encapsulating agent and bioactive compound through an orifice and formation of droplets at the discharge point of the nozzle [12]. Diverse mechanisms can be used to form droplets, such as gravitational, surface tension, impulse, and frictional forces. The extrusion techniques are classified into simple dripping, electrostatic extrusion, coaxial airflow, vibrating jet/nozzle, jet cutting, and spinning disk atomization. The droplets formed are immediately solidified to capsules by either physical or chemical process [12]. The pressure and temperature employed are typically <100 psi. In general, carbohydrate matrices have very good barrier properties and extrusion is a convenient process enabling the encapsulation of compounds such as flavours. Particles from 200 to 2000 microns have been produced by various extrusion methods with a maximal flavour load of up to 20% w/w [70].

6.3.2.4 Fluidized bed

This method was first developed by D. E. Wurster in the 1950s; hence, the term "Wurster process". Nowadays, the fluidized-bed coating method is being modified by changing the position of the nozzle to be used for coating the solid particles. The different fluidized-bed coating methods are (1) top-spray, (2) bottom-spray, and (3) tangential-spray [71]. In general, in the fluidized bed coating, granules or coated particles are produced in a single piece of equipment by spraying a binder in a solution, suspension, or melt onto a fluidized powder bed. The particles are suspended by an air stream, at a predefined temperature, and then sprayed with coating material. The evaporation of water is controlled by several factors such as spraying rate, water content, air flow, humidity of the air inlet, and temperature [12]. Typical fluidized-bed apparatus can efficiently process particles from 100 mm to a few millimetres. However, for very small particles, other forces, such as electrostatic, start to play an important role in the movement of the particles in the fluidization chamber and prevent adequate fluidization [71]. An enormous range of encapsulating agents can be used, including proteins, carbohydrates, lipids, gums and cellulose [72].

As conclusion, to select the microencapsulation technique to be used in one process, it is necessary to analyse the final product specifications, the desired particle size, release mechanisms and the physicochemical properties of the material used as the encapsulating agent and core. Also, in the case of food compounds, the cost and safety of the method also must be considered in the selection.

6.4 Encapsulating Agents: Biopolymers

The choice of an encapsulating agent is very important for the encapsulation efficiency and microparticles stability. The criteria for selecting a wall material are based on the physical-chemical properties of the substance to encapsulate (porosity, solubility) and of the encapsulating agent (viscosity, mechanical properties), the compatibility between the two (the wall material should not react with the core) and processing and economic factors [13, 54, 66, 67, 73]. Other criterion to consider is the intended size for the microcapsules.

Overall, materials for microencapsulation have to fulfil all or some of the following conditions [66, 74]:

- Have good rheological properties at high concentration (if needed) and easy workability during the encapsulation.
- Do not react with the material to be encapsulated.
- Seal and hold the active material within its structure during processing or storage.
- If applicable, completely release solvent or other materials used during encapsulation under drying or other solvent-removing conditions.
- Provide maximal protection of the core against environmental conditions.
- Be inexpensive.
- Be food-grade and legally allowed.
- Be available in large quantities and constant quality.

Several types of wall materials are available for the industry, but biopolymers are the most relevant group. The term "biopolymers" usually describes polymers produced in a natural way by living species [75]. The biopolymers should act as a barrier and it may protect the core against oxygen, water, and light or should avoid contact with other ingredients or control diffusion.

In Table 6.2, some biopolymers, suited for food microencapsulation, are presented, organized according to the main types: carbohydrate polymer, protein and lipid and the main origin: plant, marine and microbial/animal.

Table 6.2 Material suited for microencapsulation in food industry

Origin	Carbohydrate polymer	Protein	Lipid
Plant	Starch and derivatives	Gluten (corn)	Fatty acids/alcohols
	Maltodextrin	Isolates (pea, soy)	Glycerides
	Cellulose and derivatives		Phospholipids
	Plant exudates: – Gum arabic – Gum karaya – Mesquite gum		Waxes
	Plant extracts: – Galactomannans – Soluble soybean		
Marine	Carrageenan		
	Alginate		
Microbial/ animal	Xanthan	Caseins	Fatty acids/alcohols
	Gellan	Whey proteins	Glycerides
	Dextran	Gelatin	Waxes
	Chitin		Phospholipids (Shellac)
	Chitosan and derivatives		

Source: Adapted from [39].

Biopolymers can be used to create a diverse range of delivery systems suitable for encapsulating, protecting, and delivering functional components [76].

These biopolymer particles must be carefully designed and manufactured so that they exhibit the required functional attributes within the final product, such as optical properties, rheological properties, release characteristics, encapsulation properties, and physicochemical stability [38].

Therefore, the choice of a wall material is an important task. Depending on the characteristics desired in the final product, the wall material can be selected from a wide variety of biopolymers.

Different wall materials can be used and each group has certain advantages and disadvantages. In some cases, one wall material alone may not have all the required characteristics, so a combination of encapsulating agents may be used [58, 77, 78].

Two of the most important groups of biopolymers are the carbohydrates and the proteins.

6.4.1 Carbohydrates

In terms of carbohydrate polymer, some examples are very well known from the industry such as starch, cellulose, maltodextrins, gum arabic, alginates, carrageenans and chitosan. In general, the carbohydrates need to be modified or associated to other wall material [13, 67]. The ability of carbohydrates to bind compounds is complemented by their diversity, good properties, low cost and makes them the preferred choice for encapsulation, being the subject of considerable studies [58].

In the next section, the use of chitosan in microencapsulation processes is more detailed.

6.4.2 Proteins

Proteins, such as whey proteins, sodium caseinate, gelatin, whey protein and soy protein isolates, are another possibility to use as encapsulating agents [79]. Proteins are biological polymers comprising amino acids that come in a variety of different general structures: fibrous, random coil, and globular proteins. The molecular structure adopted by a protein depends on its amino acid sequence, prevailing environmental conditions, and environmental history [76].

A large number of factors must be considered when selecting an appropriate protein or combination of proteins to fabricate biopolymer-based delivery systems. It is important to know some specific physicochemical characteristics of the proteins involved, such as thermal denaturation temperatures (for globular proteins), helixecoil transition temperatures (for gelatin or collagen), isoelectric points (pI), sensitivities to specific monovalent or multivalent ions, or susceptibility to specific enzyme or chemical cross-linking or degradation reactions [76].

6.5 Microencapsulation with Chitosan

Chitosan is a widely used biopolymer. Chitosan is a polysaccharide obtained commercially by alkaline deacetylation of chitin (an N-acetyl-glucosamine polymer). Chitin is the second most abundant natural polymer in nature after cellulose and is found in the structure of a wide number of invertebrates (crustaceans, exoskeleton insects, cuticles among others) [80]; it is estimated that every year 100 billion tons of chitin are produced on the earth [40], from which chitosan is the main derivative.

The chitosan molecule is a copolymer of N-acetyl-D-glucosamine and D-glucosamine and it differs in the degree of N-acetylation (40–98%) and molecular weight (50–2000 kDa) [81]. Chitosan has interesting intrinsic properties, such as biocompatibility and biodegradability. Other chitosan's properties with special interest for the food industry are related to the anticholesterolemic, hypo-cholesterolemic, antimicrobial, and antioxidant properties [41]. So, one of the most interesting chitosan properties is the cholesterol reducing effect. There are several mechanisms studied to try to explain and describe this phenomenon [82, 83].

Chitosan is also considered as non-toxic, it has a LD50 (median lethal dose) of around 16 g/kg, comparable to the salt and glucose values [41].

Chitosan is also regarded as inexpensive product and, as it possesses reactive amino functional groups, has the potential to be used in many different fields [80, 84–87].

There are two chitosan characteristics (deacetylation degree and molecular weight) that have a significant role in the biological chitosan properties [6, 41]. The chitosan deacetylation degree is referred in literature as one of the most important characteristics involved in the chitosan characterization and the molecular weight is determinant to define the viscosity of chitosan. Some authors verify that decreasing the deacetylation degree the biodegradability increases. However, a uniform distribution of the acetylated groups contributes to decreasing the biodegradability. Another interesting property is the biocompatibility, but this property depends on the characteristics of the sample (natural source, method of preparation, molecular weight, and deacetylation degree) [41].

Chitin and chitosan are water-insoluble materials. Chitosan is soluble only in acidic solutions because of its rigid crystalline structure and deacetylation, which limit its application to bioactive agents such as drug carriers [14, 88, 89]. However, it is possible to modify the chitosan structure to form water-soluble chitosan which is easily soluble in neutral aqueous solutions. This is useful for drug carriers and for food industrial applications. In order to develop water-soluble chitosan, many studies have been carried out to modify the molecular structure of chitosan, and thereby improve or control its properties, with carboxymethyl, dihydroxyethyl, sulphuryl, or phosphoryl groups [88, 90].

Therefore, an important use of chitosan involves its application in controlled-release systems, normally as nanocapsules or microcapsules [91–95], hydrogels [6, 96], nanofibres [97], films and tablets [41, 98]. Chitosan has one important advantage over other encapsulating agents, which is the possibility to establish covalent or ionic bonds with the cross-linking agents, building a sort of network, where the active substance stays retained. Therefore, these chemical bonds carry advantages in terms of controlled release [36, 99–101]. Depending on the nature of the cross-linker, the main interactions forming the network are covalent or ionic bonds.

Covalent cross-linking leads to the formation of hydrogels or microparticles with a permanent network structure, because irreversible chemical bonds are formed. Cross-linkers agents used to establish these bonds are dialdehydes, such as glyoxal and in particular glutaraldehyde [6, 101–104]. However, the main disadvantage of this kind of cross-linking agents, dialdehydes, is that they are generally considered toxic for food application purposes. Other covalent cross-linkers have been investigated as alternatives such as diethyl squarate, oxalic acid or genipin [6, 101–105].

Ionically cross-linked microparticles or hydrogels are generally considered as biocompatible, well tolerated by the human organism, showing advantages in terms of applications for food and pharmaceutical industry [61, 100, 106]. Their network is non-permanent and forms reversible links. Ionically cross-linked chitosan hydrogels or microparticles exhibit a higher swelling sensitivity to pH changes compared to covalently cross-linked chitosan compounds. The most relevant are tripolyphosphate, citrate, sulphate and phosphate among others [61, 100, 106].

So, considering all these natural advantages, chitosan has been particularly used as support in microencapsulation processes for controlled release of bioactive compounds [25, 26, 31, 35, 47, 52, 107–112]. Thus, chitosan, in addition to having beneficial properties to health, can be considered a feasible biopolymer to be applied in various microencapsulation techniques.

6.6 Controlled-Release Studies

The concept of drug delivery has expanded into different compound delivery systems [113, 114]. With a properly designed controlled-release system, the compound is released at the desired site and time, and at a desired rate [13, 53]. The release mechanisms vary with the type and amount of the used encapsulating agent, the method of preparation, and the environment where the release occurs [48, 115].

The most common release mechanisms are processes of diffusion, biodegradation, dilatation (with gel formation and swelling) and osmosis [13, 116]. The models describing the release of the active compounds can be classified into three main categories: statistical methods (exploratory data analysis method, repeated measures design, multivariate approach [MANOVA: multivariate analysis of variance]; model dependent methods (zero order, first order, Higuchi, Korsmeyer–Peppas, Hixson–Crowell, Baker–Lonsdale, and Weibull model, among others); and model-independent methods [difference factor (f1), similarity factor (f2)] [53].

In an ideal system, the compound release may follow zero-, half- or first-order kinetics. The kinetics of zero order (constant release rate) occurs when the core is a pure material that is released from the system as a pure material. The half-order kinetics generally occurs with matrix particles. The first-order kinetics occurs when the core is actually a solution [13, 48, 53, 117–119]. Usually, the real release of the compound may be slightly different from zero, half or first order kinetics [48, 117, 118], and, because of that, more complex mathematical models attempt to describe the phenomena, namely the Higuchi and the

Korsmeyer–Peppas models. Other models have specific fields of application for example: Hixson–Crowell model is applied to pharmaceutical dosage forms such as tablets, and Weibull model is more useful for comparing the release profiles of matrix type drug delivery [53, 120]. Other models have been developed further from the main models, for example, the Baker–Lonsdale model was developed from the Higuchi model [53].

The applicability of these mathematical models to the controlled release of compounds is increasing in industry allowing the development of new systems. Sophisticated instrumentation, modern mathematical models and computational power have revolutionized the entire process of formulation and development of drug and food delivery systems.

6.7 Conclusions

This chapter presents and discusses the importance of producing microparticles with technological interest, using chitosan, showing the recent developments and the new applications of it.

The chapter starts with a brief introduction about the microencapsulation issue. After that, the main microencapsulation methods, main encapsulating agents, the application of chitosan in microencapsulation processes are discussed, and finally the mathematical modelling and evaluation of the mechanisms of controlled release are presented.

Microencapsulation has become a challenging approach to design new materials with biotechnological applications. Methods can be divided in two main types: chemical processes and mechanical processes. Most of the materials used for microencapsulation are bio-based materials such as carbohydrates, fats, waxes, and proteins. These natural polymers present several advantages such as biocompatibility, biodegradability, good amphiphilic and functional properties such as water solubility, and emulsifying capacity. Biopolymers can be used as matrixes for the microencapsulation, immobilization or controlled release of numerous active compounds and have been applied in several fields.

Chitosan is a widely used biopolymer and has interesting intrinsic properties, such as biocompatibility, biodegradability, anticholesterolemic, hypo-cholesterolemic, antimicrobial, and antioxidant properties. So, considering all these natural advantages, chitosan has been particularly used as support in microencapsulation processes for controlled release of bioactive compounds.

Finally, the release mechanisms vary with the type and amount of the used encapsulating agent, the method of preparation, and the environment where the release occurs. The most common release mechanisms are processes of diffusion, biodegradation, dilatation (with gel formation and swelling) and osmosis. Some mathematical models are also discussed.

Acknowledgements

The authors thank Fundação para a Ciência e a Tecnologia (FCT) for the grant SFRH/BPD/73865/2010 (B. N. Estevinho).

This work was also financially supported by: Project POCI-01-0145-FEDER-006939 (Laboratory for Process Engineering, Environment, Biotechnology and Energy—LEPABE funded by FEDER funds through COMPETE2020—Programa Operacional Competitividade e Internacionalização (POCI)— and by national funds through FCT—Fundação para a Ciência e a Tecnologia.

References

1. Gouin, S. (2004). Microencapsulation: Industrial appraisal of existing technologies and trends. *Trends Food Sci. Technol.* 15(7–8), pp. 330–347.
2. Champagne, C. P., and Fustier, P. (2007). Microencapsulation for the improved delivery of bioactive compounds into foods. *Curr. Opin. Biotechnol.* 18(2), pp. 184–190.
3. Estevinho, B. N., Rocha, F., Santos, L., and Alves, A. (2013). Microencapsulation with chitosan by spray drying for industry applications—A review. *Trends Food Sci. Technol.* 31(2), pp. 138–155.
4. Higuera-Ciapara, I., Felix-Valenzuela, L., Goycoolea, F. M., and Argüelles-Monal, W. (2004). Microencapsulation of astaxanthin in a chitosan matrix. *Carbohydr. Polym.* 56(1), pp. 41–45.
5. Shi, X.-Y., and Tan, T.-W. (2002). Preparation of chitosan/ethylcellulose complex microcapsule and its application in controlled release of vitamin D2. *Biomaterials* 23(23), pp. 4469–4473.

6. Berger, J., Reist, M., Mayer, J. M., Felt, O., Peppas, N. A., and Gurny, R. (2004). Structure and interactions in covalently and ionically crosslinked chitosan hydrogels for biomedical applications. *Eur. J. Pharm. Biopharm.* 57(1), pp. 19–34.
7. Blasi, P., Giovagnoli, S., Schoubben, A., Ricci, M., Rossi, C., Luca, G., Calafiore, R., et al. (2006). Preparation and in vitro and in vivo characterization of composite microcapsules for cell encapsulation. *Int. J. Pharm.* 324(1), pp. 27–36.
8. Gomaa, Y. A., El-Khordagui, L. K., Boraei, N. A., and Darwish, I. A. (2010). Chitosan microparticles incorporating a hydrophilic sunscreen agent. *Carbohydr. Polym.* 81(2), pp. 234–242.
9. Monllor, P., Bonet, M. A., and Cases, F. (2007). Characterization of the behaviour of flavour microcapsules in cotton fabrics. *Eur. Polym. J.* 43(6), pp. 2481–2490.
10. Hu, X. P., Shasha, B. S., McGuire, M. R., and Prokopy, R. J. (1998). Controlled release of sugar and toxicant from a novel device for controlling pest insects. *J. Control. Release* 50(1–3), pp. 257–265.
11. Desai, K. G. H., and Park, H. J. (2005). Recent developments in microencapsulation of food ingredients. *Dry. Technol.* 23, pp. 1361–1394.
12. Đorđević, V., Balanč, B., Belščak-Cvitanović, A., Lević, S., Trifković, K., Kalušević, A., Nedović, V., et al. (2014). Trends in encapsulation technologies for delivery of food bioactive compounds. *Food Eng. Rev.* 7(4), pp. 452–490.
13. Estevinho, B. N., Rocha, F., Santos, L., and Alves, A. (2013). Microencapsulation with chitosan by spray drying for industry applications—A review. *Trends Food Sci. Technol.* 31(2), pp. 138–155.
14. Estevinho, B. N., Rocha, F., Santos, L., and Alves, A. (2013). Using water soluble chitosan for flavour microencapsulation in food industry. *J. Microencapsul.* 30(6), pp. 571–579.
15. Roos, K. B. de (2003). Effect of texture and microstructure on flavour retention and release. *Int. Dairy J.*, 13(8), pp. 593–605.
16. Yoshii, H., Soottitantawat, A., Liu, X.-D., Atarashi, T., Furuta, T., Aishima, S., Linko, P., et al. (2001). Flavor release from spray-dried maltodextrin/gum arabic or soy matrices as a function of storage relative humidity. *Innov. Food Sci. Emerg. Technol.* 2, pp. 55–61.
17. Paramita, V., Furuta, T., and Yoshii, H. (2010). Microencapsulation efficacy of d-limonene by spray drying using various combinations of wall materials and emulsifiers. *Food Sci. Technol. Res.*, 16(5), pp. 365–372.

18. Uzun, I. (2006). Kinetics of the adsorption of reactive dyes by chitosan. *Dye. Pigment.* 70(2), pp. 76–83.
19. Ersus, S., and Yurdagel, U. (2007). Microencapsulation of anthocyanin pigments of black carrot (Daucus carota L.) by spray drier. *J. Food Eng.* 80(3), pp. 805–812.
20. Guillard, V., Issoupov, V., Redl, A., and Gontard, N. (2009). Food preservative content reduction by controlling sorbic acid release from a superficial coating. *Innov. Food Sci. Emerg. Technol.* 10(1), pp. 108–115.
21. Alvarez-parrilla, E., Rosa, L., and Torres-, F. (2005). Complexation of apple antioxidants: Chlorogenic acid, quercetin and rutin by β-cyclodextrin (β-CD)*. *J. Incl. Phenom. Macrocycl. Chem.* 53(1), pp. 121–129.
22. Casanova, F., Estevinho, B. N., and Santos, L. (2016). Preliminary studies of rosmarinic acid microencapsulation with chitosan and modified chitosan for topical delivery. *Powder Technol.* 297, pp. 44–49.
23. Casanova, F., and Santos, L. (2016). Encapsulation of cosmetic active ingredients for topical application—a review. *J. Microencapsul.* 33(1), pp. 1–17.
24. Gemili, S., Yemenicioðlu, A., and Altýnkaya, S. A. (2010). Development of antioxidant food packaging materials with controlled release properties. *J. Food Eng.* 96(3), pp. 325–332.
25. Estevinho, B. N., Ramos, I., and Rocha, F. (2015). Effect of the pH in the formation of β-galactosidase microparticles produced by a spray-drying process. *Int. J. Biol. Macromol.* 78, pp. 238–242.
26. Estevinho, B. N., Damas, A. M., Martins, P., and Rocha, F. (2014). The influence of microencapsulation with a modified chitosan (water soluble) on β-galactosidase activity. *Dry. Technol.* 32, pp. 1575–1586.
27. Estevinho, B. N., Damas, A. M., Martins, P., and Rocha, F. (2014). Microencapsulation of β-galactosidase with different biopolymers by a spray-drying process. *Food Res. Int.* 64, pp. 134–140.
28. Kishore, D., and Kayastha, A. M. (2012). Optimisation of immobilisation conditions for chick pea β-galactosidase (CpGAL) to alkylamine glass using response surface methodology and its applications in lactose hydrolysis. *Food Chem.* 134(3), pp. 1650–1657.
29. Sangeetha, K., and Abraham, T. E. (2008). Preparation and characterization of cross-linked enzyme aggregates (CLEA) of

subtilisin for controlled release applications. *Int. J. Biol. Macromol.* 43, pp. 314–319.

30. Heidebach, T., Först, P., and Kulozik, U. (2010). Influence of casein-based microencapsulation on freeze-drying and storage of probiotic cells. *J. Food Eng.* 98(3), pp. 309–316.
31. Estevinho, B. N., Carlan, I., Blaga, A., and Rocha, F. (2016). Soluble vitamins (vitamin B12 and vitamin C) microencapsulated with different biopolymers by a spray drying process. *Powder Technol.* 289, pp. 71–78.
32. Gonçalves, A., Estevinho, B. N., and Rocha, F. (2017). Design and characterization of controlled-release vitamin A microparticles prepared by a spray-drying process. *Powder Technol.* 305, pp. 411–417.
33. Gonçalves, A., Estevinho, B. N., and Rocha, F. (2016). Microencapsulation of vitamin A: A review. *Trends Food Sci. Technol.* 51, pp. 76–87.
34. Yoo, S.-H., Song, Y.-B., Chang, P.-S., and Lee, H. G. (2006). Microencapsulation of alpha-tocopherol using sodium alginate and its controlled release properties. *Int. J. Biol. Macromol.* 38(1), pp. 25–30.
35. Comunian, T. A., and Favaro-Trindade, C. S. (2016). Microencapsulation using biopolymers as an alternative to produce food enhanced with phytosterols and omega-3 fatty acids: A review. *Food Hydrocoll.* 61, pp. 442–457.
36. Estevinho, B. M. A. N., Rocha, F. A. N., Santos, L. M. D. S., and Alves, M. A. C. (2013). Using water-soluble chitosan for flavour microencapsulation in food industry. *J. Microencapsul.* 30(6), pp. 571–579.
37. Nesterenko, A., Alric, I., and Durrieu, V. (2013). Vegetable proteins in microencapsulation: A review of recent interventions and their effectiveness, 42, pp. 469–479.
38. Jones, O. G., and McClements, D. J. (2010). Functional biopolymer particles: Design, fabrication, and applications. *Compr. Rev. Food Sci. Food Saf.* 9(4), pp. 374–397.
39. Wandrey, C., Bartkowiak, A., and Harding, S. E. (2010). Materials for encapsulation. *Encapsulation Technologies for Active Food Ingredients and Food Processing,* pp. 31–100.
40. Rege, P. R., Garmise, R. J., and Block, L. H. (2003). Spray-dried chitinosans. Part I: Preparation and characterization. *Int. J. Pharm.* 252(1–2), pp. 41–51.

41. Aranaz, I., Mengíbar, M., Harris, R., Paños, I., Miralles, B., Acosta, N., Heras, Á, et al. (2009). Functional characterization of chitin and chitosan. *Curr. Chem. Biol.* 3, pp. 203–230.

42. Alexakis, T., Boadi, D. K., Quong, D., Groboillot, A., O'Neill, I., Poncelet, D., and Neufeld, R. J. (1995). Microencapsulation of DNA within alginate microspheres and crosslinked chitosan membranes for in vivo application. *Appl. Biochem. Biotechnol.* 50(1), pp. 93–106.

43. Jameela, S. R., and Jayakrishnan, A. (1995). Glutaraldehyde cross-linked chitosan microspheres as a long acting biodegradable drug delivery vehicle: Studies on the in vitro release of mitoxantrone and in viva degradation of microspheres in rat muscle. *Biomaterials*, 16(10), pp. 769–775.

44. González Siso, M. I., Lang, E., Carrenõ-Gómez, B., Becerra, M., Otero Espinar, F., and Blanco Méndez, J. (1997). Enzyme encapsulation on chitosan microbeads. *Process Biochem.* 32(3), pp. 211–216.

45. Estevinho, B. N., Damas, A. M., Martins, P., and Rocha, F. (2012). Study of the inhibition effect on the microencapsulated enzyme β-galactosidase. *Environ. Eng. Manag. J.* 11(11), pp. 1923–1930.

46. Estevinho, B. N., Damas, A. M., Martins, P., and Rocha, F. (2014). The influence of microencapsulation with a modified chitosan (water soluble) on β-galactosidase activity. *Dry. Technol.* 32, pp. 1575–1586.

47. Ko, J. A., Park, H. J., Park, Y. S., Hwang, S. J., and Park, J. B. (2003). Chitosan microparticle preparation for controlled drug release by response surface methodology. *J. Microencapsul.* 20(6), pp. 791–797.

48. Pothakamury, U. R., and Barbosa-Cánovas, G. V. (1995). Fundamental aspects of controlled release in foods. *Trends Food Sci. Technol.* 6 (December), pp. 397–406.

49. Carvalho, I. T., Estevinho, B. N., and Santos, L. (2016). Application of microencapsulated essential oils in cosmetic and personal healthcare products—A review. *Int. J. Cosmet. Sci.* 38(2), pp. 109–119.

50. Estevinho, B. N., and Rocha, F. (2016). Microencapsulation in food biotechnology by a spray-drying process. In Ravishankar Rai V. (ed.), *Advances in Food Biotechnology* (pp. 593–606). John Wiley and Sons, Ltd.

51. Carvalho, I. T., Estevinho, B. N., and Santos, L. (2016). Application of microencapsulated essential oils in cosmetic and personal health care products—A review. *Int. J. Cosmet. Sci.* 38, pp. 109–119.

52. Casanova, F., Estevinho, B. N., and Santos, L. (2016). Preliminary studies of rosmarinic acid microencapsulation with chitosan

and modified chitosan for topical delivery. *Powder Technol.* 297, pp. 44–49.

53. Dash, S., Murthy, P. N., Nath, L., and Chowdhury, P. (2010). Kinetic modeling on drug release from controlled drug delivery systems. *Acta Pol. Pharm. Drug Res.* 67(3), pp. 217–223.
54. Ghosh, S. K. (2006). Functional coatings and microencapsulation: a general perspective. In S. K. Ghosh (ed.), *Functional Coatings* (pp. 1–28). Weinheim: WILEY-VCH Verlag GmbH and CO KGaA.
55. Abbas, S., Da Wei, C., Hayat, K., and Xiaoming, Z. (2012). Ascorbic acid: Microencapsulation techniques and trends—a review. *Food Rev. Int.* 28(4), pp. 343–374.
56. Gonçalves, A., Estevinho, B. N., and Rocha, F. (2016). Microencapsulation of vitamin A: A review. *Trends Food Sci. Technol.* 51, pp. 76–87.
57. Vos, P. de, Faas, M. M., Spasojevic, M., and Sikkema, J. (2010). Encapsulation for preservation of functionality and targeted delivery of bioactive food components. *Int. Dairy J.* 20(4), pp. 292–302.
58. Madene, A., Jacquot, M., Scher, J., and Desobry, S. (2006). Flavour encapsulation and controlled release—a review. *Int. J. Food Sci. Technol.* 41(1), pp. 1–21.
59. Eraso, M. O., and Herrera, A. (2013). Use of starches and milk proteins in microencapsulation. *Int. J. Veg. Sci.* (January 2014), pp. 1–29.
60. Kaushik, P., Dowling, K., Barrow, C. J., and Adhikari, B. (2015). Microencapsulation of omega-3 fatty acids: A review of microencapsulation and characterization methods. *J. Funct. Foods.* 19, pp. 868–881.
61. Desai, K. G. H., and Park, H. J. (2005). Encapsulation of vitamin C in tripolyphosphate cross-linked chitosan microspheres by spray drying. *J. Microencapsul.* 22(2), pp. 179–192.
62. Pu, J., Bankston, J. D., and Sathivel, S. (2011). Developing microencapsulated flaxseed oil containing shrimp (Litopenaeus setiferus) astaxanthin using a pilot scale spray dryer. *Biosyst. Eng.* 108(2), pp. 121–132.
63. Schafroth, N., Arpagaus, C., Jadhav, U. Y., Makne, S., and Douroumis, D. (2011). Nano and microparticle engineering of water insoluble drugs using a novel spray–drying process. *Colloids Surf. B Biointerfaces.* 90, pp. 8–15.

64. Rattes, A. L. R., and Oliveira, W. P. (2007). Spray drying conditions and encapsulating composition effects on formation and properties of sodium diclofenac microparticles. *Powder Technol.* 171(1), pp. 7–14.

65. Jafari, S. M., Assadpoor, E., He, Y., and Bhandari, B. (2008). Encapsulation efficiency of food flavours and oils during spray drying. *Dry. Technol.* 26(7), pp. 816–835.

66. Azeredo, H. M. C. de. (2005). Encapsulação: Aplicação à tecnologia de alimentos. *Alim. Nutr. Araraquara* 16(1), pp. 89–97.

67. Gharsallaoui, A., Roudaut, G., Chambin, O., Voilley, A., and Saurel, R. (2007). Applications of spray-drying in microencapsulation of food ingredients: An overview. *Food Res. Int.* 40(9), pp. 1107–1121.

68. Sillick, M., and Gregson, C. M. (2012). Spray chill encapsulation of flavors within anhydrous erythritol crystals. *LWT—Food Sci. Technol.* 48(1), pp. 107–113.

69. Palzer, S. (2009). Food structures for nutrition, health and wellness. *Trends Food Sci. Technol.* 20(5), pp. 194–200.

70. Manojlovic, V., Rajic, N., Djonlagic, J., Obradovic, B., Nedovic, V., and Bugarski, B. (2008). Application of electrostatic extrusion–flavour encapsulation and controlled release. *Sensors,* 8(3), pp. 1488–1496.

71. Desai, K. G. H., and Jin Park, H. (2005). Recent developments in microencapsulation of food ingredients. *Dry. Technol.* 23(7), pp. 11361–1394.

72. Sun, P., Zeng, M., He, Z., Qin, F., and Chen, J. (2013). Controlled release of fluidized bed-coated menthol powder with a gelatin coating. *Dry. Technol.* 31(13–14), pp. 1619–1626.

73. Freiberg, S., and Zhu, X. X. (2004). Polymer microspheres for controlled drug release. *Int. J. Pharm.* 282(1–2), pp. 1–18.

74. Mahdavi, S. A., Jafari, S. M., Ghorbani, M., Mahdavi, S. A., Jafari, S. M., and Ghorbani, M. (2014). Spray-drying microencapsulation of anthocyanins by natural biopolymers: A review. *Dry. Technol.* 32(5), pp. 509–518.

75. Chassenieux, C., Durand, D., Jyotishkumar, P., and Thomas, S. (2013). Biopolymers: State of the art, new challenges, and opportunities. *Handbook of Biopolymers-Based Materials: From Blends and Composites to Gels and Complex Networks,* pp. 1–6.

76. Matalanis, A., Jones, O. G., and McClements, D. J. (2011). Structured biopolymer-based delivery systems for encapsulation, protection, and release of lipophilic compounds. *Food Hydrocoll.* 25(8), pp. 1865–1880.

77. Cortesi, R., Nastruzzi, C., and Davis, S. S. (1998). Sugar cross-linked gelatin for controlled release: Microspheres and disks. *Biomaterials*, 19(18), pp. 1641–1649.

78. Favaro-Trindade, C. S., Santana, A. S., Monterrey-Quintero, E. S., Trindade, M. A., and Netto, F. M. (2010). The use of spray drying technology to reduce bitter taste of casein hydrolysate. *Food Hydrocoll.* 24(4), pp. 336–340.

79. Aghbashlo, M., Mobli, H., Rafiee, S., and Madadlou, A. (2012). Energy and exergy analyses of the spray drying process of fish oil microencapsulation. *Biosyst. Eng.* 111(2), pp. 229–241.

80. Kumar, M. N. V. R. (2000). A review of chitin and chitosan applications. *React. Funct. Polym.* 46, pp. 1–27.

81. Guliyeva, U., Oner, F., Ozsoy, S., and Haziroðlu, R. (2006). Chitosan microparticles containing plasmid DNA as potential oral gene delivery system. *Eur. J. Pharm. Biopharm.* 62(1), pp. 17–25.

82. Muzzarelli, R. A. A., Orlandini, F., Pacetti, D., Boselli, E., Frega, N. G., Tosi, G., and Muzzarelli, C. (2006). Chitosan taurocholate capacity to bind lipids and to undergo enzymatic hydrolysis: An in vitro model. *Carbohydr. Polym.* 66(3), pp. 363–371.

83. Zhou, K., Xia, W., Zhang, C., and (Lucy) Yu, L. (2006). In vitro binding of bile acids and triglycerides by selected chitosan preparations and their physico-chemical properties. *LWT–Food Sci. Technol.* 39(10), pp. 1087–1092.

84. Bodmeier, R. (1997). Tableting of coated pellets. *Eur. J. Pharm. Biopharm.* 43, pp. 1–8.

85. Wu, H., Wang, S., Fang, H., Zan, X., Zhang, J., and Wan, Y. (2011). Chitosan-polycaprolactone copolymer microspheres for transforming growth factor-β1 delivery. *Colloids Surf. B. Biointerfaces.* 82(2), pp. 602–608.

86. Estevinho, B. N., Ferraz, A, Rocha, F., Alves, A, and Santos, L. (2008). Interference of chitosan in glucose analysis by high-performance liquid chromatography with evaporative light scattering detection. *Anal. Bioanal. Chem.* 391(4), pp. 1183–1188.

87. Estevinho, B. N., Ferraz, A., Rocha, F., Santos, L., and Alves, A. (2009). Uncertainty in the determination of glucose in aqueous solutions by high-performance liquid chromatography with evaporative light scattering detection. *J. Sep. Sci.* 32(18), pp. 3116–3125.

88. Sashiwa, H., Kawasaki, N., and Nakayama, A. (2002). Chemical modification of chitosan. 14:1 Synthesis of water-soluble chitosan

derivatives by simple acetylation. *Biomacromolecules,* 3(5), pp. 1126–1128.

89. Zhang, H., Wu, S., Tao, Y., Zang, L., and Su, Z. (2010). Preparation and characterization of water-soluble chitosan nanoparticles as protein delivery system. *J. Nanomater.* 2010, pp. 1–5.
90. Chung, Y. C., Tsai, C. F., and Li, C. F. (2006). Preparation and characterization of water-soluble chitosan produced by Maillard reaction. *Fish. Sci.* 72(5), pp. 1096–1103.
91. Keegan, G. M., Smart, J. D., Ingram, M. J., Barnes, L.-M., Burnett, G. R., and Rees, G. D. (2012). Chitosan microparticles for the controlled delivery of fluoride. *J. dent.* 40(3), pp. 229–240.
92. Ee, H. Y. G. Y. U. L. (2008). Stability of chitosan nanoparticles for L-ascorbic. *System*, pp. 1936–1941.
93. Wang, C., Fu, X., and Yang, L. (2007). Water-soluble chitosan nanoparticles as a novel carrier system for protein delivery. *Chinese Sci. Bull.* 52(7), pp. 883–889.
94. Bastos, D., da, S., Gonçalves, M., do, P., Andrade, C. T. de, Araújo, K., de, L. G., and Leão, M. H. M., da, R. (2012). Microencapsulation of cashew apple (Anacardium occidentale, L) juice using a new chitosan-commercial bovine whey protein isolate system in spray drying. *Food Bioprod. Process.* 90, pp. 683–692.
95. Liu, C., Desai, K. G. H., Tang, X., Chen, X., Liu, C., Desai, K. G. H., Chen, X., et al. (2006). Drug release kinetics of spray-dried chitosan microspheres drug release kinetics of spray-dried chitosan microspheres. *Dry. Technol. ISSN* 24, pp. 769–776.
96. George, M., and Abraham, T. E. (2006). Polyionic hydrocolloids for the intestinal delivery of protein drugs: Alginate and chitosan—a review. *J. Control. Release.* 114(1), pp. 1–14.
97. Bhattarai, N., Edmondson, D., Veiseh, O., Matsen, F. A., and Zhang, M. (2005). Electrospun chitosan-based nanofibers and their cellular compatibility. *Biomaterials*, 26(31), pp. 6176–6184.
98. Felt, O., Buri, P., and Gurny, R. (1998). Chitosan: A unique polysaccharide for drug delivery. *Drug Dev. Ind. Pharm.* 24(11), pp. 979–993.
99. Desai, K. G. H., and Park, H. J. (2005). Preparation of cross-linked chitosan microspheres by spray drying: Effect of cross-linking agent on the properties of spray dried microspheres. *J. Microencapsul.* 22(4), pp. 377–395.
100. Shu, X. Z., and Zhu, K. J. (2000). A novel approach to prepare tripolyphosphate/chitosan complex beads for controlled release drug delivery. *Int. J. Pharm.* 201(1), pp. 51–58.

101. Wang, L.-Y., Ma, G.-H., and Su, Z.-G. (2005). Preparation of uniform sized chitosan microspheres by membrane emulsification technique and application as a carrier of protein drug. *J. Control. Release.* 106(1–2), pp. 62–75.

102. Ganza-González, A., Anguiano-Igea, S., Otero-Espinar, F. J., and Méndez, J. B. (1999). Chitosan and chondroitin microspheres for oral-administration controlled release of metoclopramide. *Eur. J. Pharm. Biopharm.* 48(2), pp. 149–155.

103. Genta, I., Perugini, P., Conti, B., and Pavanetto, F. (1997). A multiple emulsion method to entrap a lipophilic compound into chitosan microspheres. *Int. J. Pharm.* pp. 237–246.

104. Muzzarelli, R. A. A. (2009). Genipin-crosslinked chitosan hydrogels as biomedical and pharmaceutical aids. *Carbohydr. Polym.* 77(1), pp. 1–9.

105. Chen, H., Ouyang, W., Martoni, C., and Prakash, S. (2009). Genipin cross-linked polymeric alginate-chitosan microcapsules for oral delivery: In-vitro analysis. *Int. J. Polym. Sci.* 2009, pp. 49–52.

106. Shu, X. Z., and Zhu, K. J. (2002). Controlled drug release properties of ionically cross-linked chitosan beads: The influence of anion structure. *Int. J. Pharm.* 233(1–2), pp. 217–225.

107. Estevinho, B. N., Damas, A. M., Martins, P., and Rocha, F. (2014). Microencapsulation of β-galactosidase with different biopolymers by a spray-drying process. *Food Res. Int.* 64, 134–140.

108. Estevinho, B. N., Rocha, F., Santos, L., and Alves, M. (2013). Using water-soluble chitosan for flavour microencapsulation in food industry. *J. Microencapsul.* 30(6), pp. 571–579.

109. Solanki, H. K., Pawar, D. D., Shah, D. A., Prajapati, V. D., Jani, G. K., Mulla, A. M., and Thakar, P. M. (2013). Development of microencapsulation delivery system for long-term preservation of probiotics as biotherapeutics agent. *Biomed Res. Int.* 2013 (2013), Article ID 620719, 21 pageshttp://dx.doi.org/10.1155/2013/620719.

110. Zhang, S., Chen, J., Yin, X., Wang, X., Qiu, B., Zhu, L., and Lin, Q. (2017). Microencapsulation of tea tree oil by spray-drying with methyl cellulose as the emulsifier and wall material together with chitosan/alginate. *J. Appl. Polym. Sci.* 134(13), pp. 1–10.

111. Xiao, Z., Liu, W., Zhu, G., Zhou, R., and Niu, Y. (2014). A review of the preparation and application of flavour and essential oils microcapsules based on complex coacervation technology. *J. Sci. Food Agric.* 94(8), pp. 1482–1494.

112. Xiao, Z., Liu, W., Zhu, G., Zhou, R., and Niu, Y. (2014). Production and characterization of multinuclear microcapsules encapsulating lavender oil by complex coacervation. *Flavour Fragr. J.* 29(3), pp. 166–172.

113. Patel, R. R., and Patel, J. K. (2010). Novel technologies of oral controlled release drug delivery system. *Syst. Rev. Pharm.* 1(2), pp. 128–132.

114. Huang, X., and Brazel, C. S. (2001). On the importance and mechanisms of burst release in matrix-controlled drug delivery systems. *J. Control. Release.* 73(2–3), pp. 121–136.

115. Ito, R., Golman, B., and Shinohara, K. (2003). Controlled release with coating layer of permeable particles. *J. Control. Release.* 92(3), pp. 361–368.

116. Lam, P. L., and Gambari, R. (2014). Advanced progress of microencapsulation technologies: In vivo and in vitro models for studying oral and transdermal drug deliveries. *J. Control. Release.* 178, pp. 25–45.

117. Sankalia, M. G., Mashru, R. C., Sankalia, J. M., and Sutariya, V. B. (2007). Reversed chitosan-alginate polyelectrolyte complex for stability improvement of alpha-amylase: Optimization and physicochemical characterization. *Eur. J. Pharm. Biopharm.* 65, pp. 215–232.

118. Sood, A., and Panchagnula, R. (2003). Design of controlled release delivery systems using a modified pharmacokinetic approach: A case study for drugs having a short elimination half-life and a narrow therapeutic index. *Int. J. Pharm.* 261(1–2), pp. 27–41.

119. Guo, W., Quan, P., Fang, L., and Cun, D. (2015). Sustained release donepezil loaded PLGA microspheres for injection: Preparation, in vitro and in vivo study. *Asian J. Pharm. Sci.* 10(5), pp. 405–414.

120. Antal, I., Zelkó, R., Roczey, N., Plachy, J., and Rácz, I. (1997). Dissolution and diffuse reflectance characteristics of coated theophylline particles. *Int. J. Pharm.* 155, pp. 83–89.

Chapter 7

Current Innovative Chitosan-Based Water Treatment of Heavy Metals: A Sustainable Approach

Tanvir Arfin
Environmental Materials Division, CSIR-NEERI, Nehru Marg, Nagpur, Maharashtra, India
tanvirarfin@gmail.com

The population in the past century quadrupled along with the sevenfold increase in the demand for water. Global water pollution has now become the ticking time bomb and requires an efficient set of technologies, policies, and scientific discoveries on incredibly diverse scales. The global water scenario will become more worrisome as the population of developing countries grows at a fast pace in the coming years. The world population is expected to rise nearly 40%, and thus the pressure on water resources caused by demand from agriculture, industry, domestic, and energy sectors will continue to grow. This chapter discusses the current research in innovative and emerging water treatment technologies based on chitin and chitosan. A number of methods that have

Handbook of Biopolymers: Advances and Multifaceted Applications
Edited by Shakeel Ahmed, Suvardhan Kanchi, and Gopalakrishnan Kumar

ISBN 978-981-4800-17-4 (Hardcover), 978-0-429-02475-7 (eBook)
www.panstanford.com

been adopted to remove toxic pollutants from contaminated water comprise ion exchange, filtration, technologies related to the membrane, chemical precipitation, and recovery of evaporation. However, such techniques are sometimes inefficient, or sometimes they cannot be regulated effectively. To keep in mind, here we focus on the chitosan-based emerging technology for efficient water purification. Currently, the application of chitosan as an adsorbent has fascinated awareness in water treatment owing to its high content of hydroxyl and amino functional groups. This miraculous polysaccharide is the deacetylated form of chitin and has a high potential for the adsorption of metal ions, dyes, and proteins, and hence is the best candidate for releasing pollutants from water and wastewater. The global scientific approach of this chitosan-based technology is being suggested to meet the needs of the current sustainable development goal.

7.1 Introduction

Water is essential for all living organisms on the earth. Life without water could not have been possible at all. Space programs are making efforts to search life on another planet, possible through the presence of water. It is necessary to investigate what percentage of water is available on the earth and the amount of water polluted through different sources and also to purify it for consumption.

The water that does not contain pathogens or toxic substances is regarded as safe for human beings. Pure water is an essential feedstock for various industries, namely drug, agriculture, etc. [1]. With such increase in population, the demand and needs of people are also growing rapidly, which has resulted in problems such as water pollution [2], the release of solid waste [3], global warming [4], and environmental degradation [5]. This has a negative impact on economic and community development. Hence, to overcome the problems caused by water pollution and to attain better eco-friendly sustainability, new and innovative techniques should be used. The access to safe drinking water is necessary for human health and better functioning of society. According to the WHO,

1.1 billion people in world do not get safe drinking water, whereas, 2.4 billion are devoid of basic sanitation facility.

7.2 Sources of Contamination

The sources of drinking water include aquifers and wells. These sources are continuously in contact with agents that lead to contamination. Such agents of contamination are shown in Fig. 7.1.

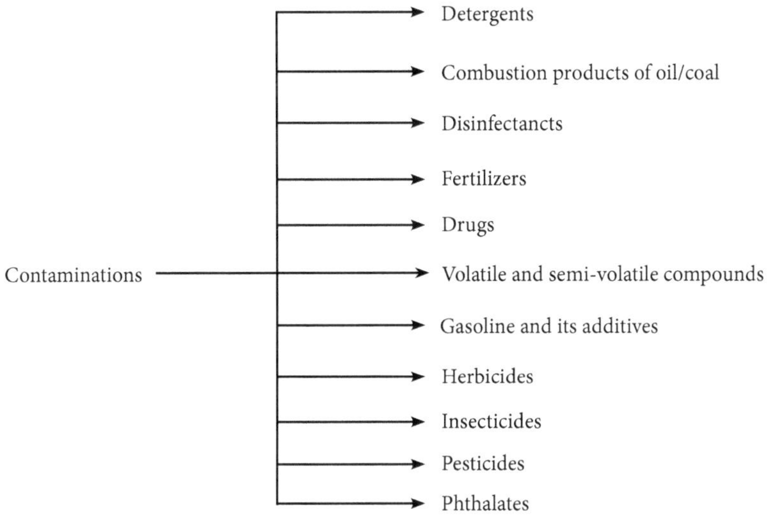

Figure 7.1 Sources of contamination.

7.3 Chitosan

Chitosan is a biopolymer material that has attracted scientists' attention in the past few years owing to its various features such as low cost and availability of amino and hydroxyl groups, which aid in the removal of the polluting agent from water [6]. It is a natural polysaccharide produced by either partial or complete deacetylation of chitin molecule. Chitin is a naturally found polysaccharide next to cellulose and serves as an important source of chitosan. It is the main constituent present in the cell wall of various fungi and the exoskeleton of crustaceans and insects present in water [7].

Chitosan is a derivative of chitin in the N-deacetylated form where the linear as well as the semicrystalline polysaccharide is made up of glucosamine, and N-acetyl glucosamine component is linked together with the help of by β-(1→4) glycosidic bonds, as shown in Fig. 7.2 [8].

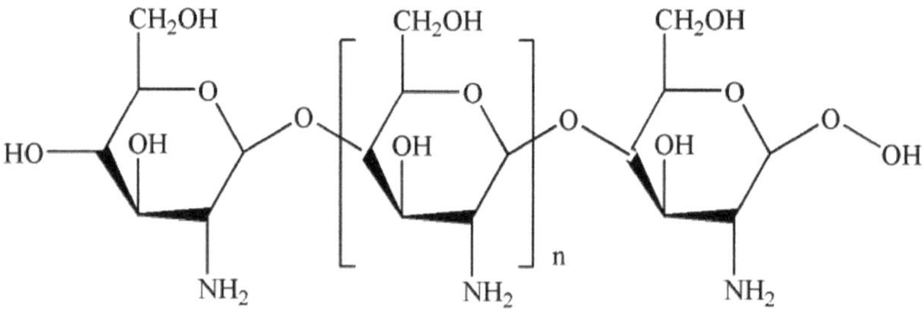

Figure 7.2 Chitosan structure.

Compared with other biomaterials, chitosan possesses exclusive properties. Namely, it is highly reactive and possesses unusual chelation behavior, steadiness, and high affinity for pollutants, which makes it attractive for research work. The natural chitosan is altered using different modes or by hybrids to enhance the adsorption process for pollutants.

Owing to numerous features such as quantum size and small size, chitosan-based materials are regarded as a suitable material for treating water. It is superior to micro-sized chitosan, as it possesses rapid adsorption capacities and rates [9].

7.4 Heavy-Metal Pollution

Heavy metals exist in the form of natural constituents in the earth's crust as well as in rocks and soils. Since no particular diagnosis of heavy metal has been providing, but it is regarded as underground metal or metalloid with atomic weight ranging from 63.5 to 200.6, specific gravity more than 5 g/ml [10]. There are various toxic metals in the water such as arsenic and mercury [11].

7.5 Conventional Procedures

Various methods have been applied for the removal of heavy metals from drinking water, including conventional methods, as shown in Fig. 7.3.

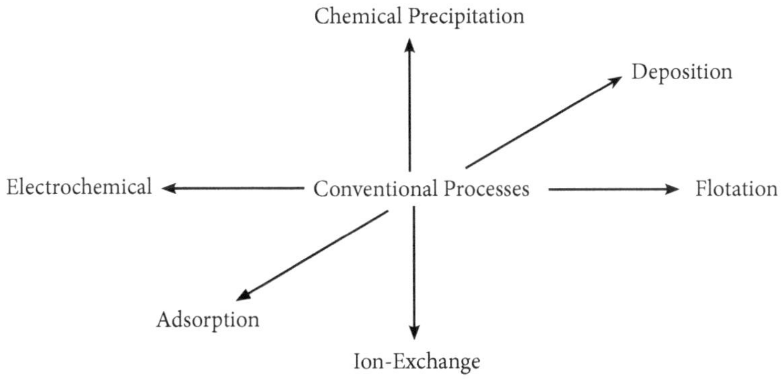

Figure 7.3 Various conventional procedures for heavy metal removal from water.

7.6 Heavy Metal Removal

Removal of heavy metals from drinking water is of utmost importance for health. Chitosan is a promising material for this purpose owing to its unique features, namely the availability of NH_2 and –OH functional groups, non-toxic effect, least cost, and availability in sufficient quantity.

7.6.1 Iron

Iron is the fourth most abundant element in the earth's crust and is in significant quantities in the biosphere as well [12]. It is mostly present in waste material released from coal coking, steel tempering, etc. [13]. The excess amount of iron particles in water leads to water pollution. The amount of iron in drinking water estimated in China is 0.3 mg/L, which even changes the taste of water. The high level of Fe(III) ions in blood is toxic for tissues as well as the different organs of the body [14]. Hence, it is the first important task to remove iron ions from water.

Wang et al. (2008) synthesized carboxymethyl chitosan (CMCts) hydrogel through γ-ray irradiation in a certain condition where the process was rapid and followed Langmuir isotherm kinetic model. The adsorption process was very fast, corresponded well to the Langmuir isotherm, and the highest acceptance of Fe(III) ion was obtained to be 18.5 mg/g gel in the adsorption process [15].

7.6.2 Copper

Copper is the contaminated substance present in the liver and the gut. Since it is a contaminant, the processing technique or tools applied pollutes the food and water also [16]. Copper, if inhaled in excess amounts, may accumulate in the liver. It is toxic to aquatic life even if it is present in small concentrations.

Sun and Wang (2006) developed N,O-CMC material and used it successfully to analyze the absorption capability on Cu^{2+} ion in aqueous solution. The results showed that it followed the Langmuir isotherm and it fitted well by the second-order reaction. It was clear that the adsorption capacity was dependent on various factors such as the degree of substitution (DS) and ionic strength. The obtained data could be later applied for framing a system to treat waste effluents [17].

7.6.3 Cadmium

Cadmium is the primary source of endemic sickness affecting specific race of people in the developing countries. It is found in the earth's crust combined with various elements, for example, with oxygen as CdO, with chlorine as $CdCl_2$, and with sulfur as CdS. It is released to the surface water from the industrial waste, which makes the water unsafe for drinking. Cadmium enters the body via drinking, breathing, or eating. It accumulates in the kidney and the liver, which leads to severe damage and causes various diseases such as emphysema and Itai-itai [18]. Several industries are responsible for releasing cadmium, namely pigments, cadmium-nickel batteries, and alloys.

Salah et al. (2013) prepared nHAp and nHApCs material for removing cadmium from water. Various parameters were used to evaluate the uptake of Cd^{2+} such as pseudo second-order equation and Langmuir and Freundlich models. They were capable of removing around 92% of cadmium. The sorption capacity of nHAp and nHApCs over cadmium was noted to be 92 and 122 mg/g, which was regarded as best material compared with the other one. The regeneration efficacy attained for nHAp when treated with cadmium was 60% in the presence of slight acidic medium [19].

7.6.4 Nickel

Nickel is the metal released during electroplating processes. When ingested, it gets absorbed and is toxic to the neural system. It even leads to gene mutations. Hence, it is the principal challenge to remove nickel from waste streams.

Wang and Kuo (2008) prepared CS nanoparticles of various sizes ranging 100–195 nm using the ionic gelation of PAA and CS. Various parameters of CS-PAA for Ni^{2+} were estimated. The monolayer capacity resulted as 435 mg/g. It was noticed that it fitted well with pseudo second order [20].

7.6.5 Chromium

Chromium(VI) is a toxic pollutant owing to its teratogenic and carcinogenic nature. The WHO has limited the amount of Cr in water at 0.05 mg/L [21]. It is discharged from industries in the form of steel and metal-finishing [22]. In water, Cr(VI) acts as a strong oxidant in the form of oxyanions such as $Cr_2O_7^{2-}$ and CrO_4^{2-} [23].

Kousalya et al. (2010) prepared n-HAp along with CS for discarding Cr from the water. The absorption capacity was 3450 mg/Kg. Since the process carried was endothermic and spontaneous in nature. It showed second-order reaction taking place by intra-particle diffusion. The material possesses properties such as biodegradability and biocompatibility. It also yields material that could be further applied in the adsorption method for wastewater management [24].

7.6.6 Cobalt

Cobalt is a heavy metal found in the nuclear waste water and leads to severe hazardous effects. The intake of cobalt may cause various health-related problems such as heart dilation and vomiting. The isotopes of cobalt liberate ionizing radiation, which may lead to cancer formation. Hence, it is necessary to undertake the issue regarding cobalt recovery and removal from various sources. Various industrial sectors such as paints and mining [25] release compounds containing cobalt. If the amount of cobalt is more in water, it can cause certain diseases such as diarrhea and bone defects. The limit of cobalt in water is fixed at 0.05 mg/L [26].

Chang *et al.* (2006) synthesized Fe_3O_4-CMCH through carbodiimide activation, where the size of the particle was taken as 13.5 nm. The material could actually remove Co(II) ion at pH 3–7 and temperature ranging 20–45°C. The adsorption followed Langmuir model capacity, which was 27.5 mg g^{-1}. In the unavailability of internal diffusion resistance, the rapid adsorption mechanism was achieved at a large surface area of nano-adsorbent. It was concluded that the electrostatic interaction between Co(II) ion and chitosan was more at low temperatures [27].

7.6.7 Arsenic

Arsenic is found in both aquatic and terrestrial environments owing to the weathering of rocks and soil destruction, formation of volcanoes. It is also found in high concentrations with the preciously deposited metals as well. It occurs naturally in various oxidation states and appears as organic and inorganic forms in water. The inorganic forms of arsenic, namely As(III) and As(IV), are toxic and cause harmful diseases such as CNS and cardiovascular disorders [28].

Santos et al. (2011) developed CS-Fe compound for the removal of As(III) from water. Its capacity was 13.4 mg/g. The kinetic study revealed that it fitted well with first order and the reaction was fast. Mossbauer spectroscopy made it clear that there was no redox reaction on the absorbents surface [29].

7.6.8 Mercury

Mercury is considered an abundant element that is toxic to the various layer of the atmosphere. The toxic nature of Hg is due to the redox reaction [30]. The highly reactive Hg(II) is too toxic, binding the amino acid cysteine within the protein molecule. The transport mechanism is dangerous in case of the elemental mercury and the organo-mercury compound. Mercury, when inhaled in vapor form, goes through the bloodstream in the lungs and spreads in the entire body system. There are many ill effects of mercury as a poisonous substance, such as neuronal disorders and damage to the cardiovascular system, liver, kidney, etc. Mercury is emitted in the body as three sources, namely natural, anthropogenic and re-emission. The primary anthropogenic sources of releasing Hg into the environment are discharges from urban areas and materials obtained from agriculture, mining, and industrial waste [31]. For minimizing the harmful effects caused by mercury, the first task should be to isolate it and then formulate a good technique to get rid of it.

Zhou et al. (2010) successfully synthesized EMCR material. Data were found by the Langmuir isotherms model, where the value of adsorption capacity was 2.69 mmol/g for mercury. During the experiment, thermodynamic and kinetic values were also calculated. The obtained value showed that the process of adsorption was exothermic in nature and it favored pseudo-second-order kinetic model. The material was regenerated at more than 90% efficiency on applying the range of solution from 0.02 M to 0.1 M of thiourea. It exhibited excellent durability with natural regeneration. The material has the ability of water uptake, which helps in the removal of Hg in drinking water as well as wastewater [32].

7.6.9 Lead

Lead is a considered a very dangerous heavy metal [33]. It is non-biodegradable and accumulates in living cells leading to various infections [34]. Pb is present in various active species. The main source of Pb is corrosion in metallic pipes [35, 36]. The availability of Pb can cause disorders of the CNS and the kidney system [37, 38].

Liu et al. (2013) suggested the adsorption features of Pb-IDMCM for lead ions in a different situation. For evaluating the adsorption properties, selectivity, and reusability, batch absorption studies were carried out. The value of adsorption capacity was 359.68 mg/g at pH 6 and 30°C. Pb-IDMCM showed high stability and proved that it is a good biosorbent for removing lead from aqueous solutions [39].

7.6.10 Zinc

Zinc is found in traces and is necessary for health in all organisms. However, when consumed in excess or insufficient amounts, it may lead to various diseases. The toxic effect of Zn in human beings takes place in either the acute or the chronic way. The excessive intake of Zn causes health problems such as diarrhea and vomiting. Zn and its salts have toxic effects on the marine life. Zn serves various purposes, namely anti-rust coatings and castings. As Zn is produced in high amounts, it results in emission sludge and effluents.

Kyzas et al. (2015) synthesized a compound called CSUC by modifying chitosan from grafting the COOH groups of succinic anhydride on the backbone of CSUC. It was found to be promising for removing zinc from wastewater [40].

7.6.11 Silver

Silver is widely used for catalysis, etc. Ag nanoparticles and ions are used as disinfectants [41] and in pharmaceuticals as well [42]. Silver ions [AgI or Ag+], if present in drinking water, are dangerous for health [43, 44]. The WHO and USEPA have limited the concentration of Ag in drinking water at 0.1 mg/L [45].

Song et al. (2012) prepared a novel material called chitosan hydrogen imprinted through silver. Different amounts of silver were taken for the experiments in a wide concentration range. The performance of silver was noted because of the CS matrix made with imprinted silver ions. It was observed that at approximately pH 5, the metal was quickly recovered by several washings through acid [46].

7.6.12 Manganese

Manganese is a heavy metal present in many sources of water. It contaminates water in either regular or manmade form, which results from the mineral dissolution from rocks [47]. It may be due to contamination coming from MnO_2 treatment, steel blend dry cells, etc. [48]. It is non-biodegradable and leads to several disorders [49].

Al-Wakeel et al. (2015) synthesized G@Chs for the removal of manganese ions from an aqueous solution. The material showed adsorption capacity to be 1.3 mmol/g at pH 6 and 25°C. The regeneration process was performed thrice, and an efficiency of 96.4%, 94.1%, and 93.5% was observed. This shows that the resin has great sturdiness and additionally exceptional potential for utilization [50].

7.7 Conclusion

This chapter focused on the removal of heavy metals from drinking water and wastewater using chitosan nanomaterials. It showed the capability of nanomaterials to remove various heavy metal ions from aqueous solutions. There are many feature of nanomaterials that makes them suitable for removing heavy metal ions, such as (a) high surface-area-to-volume ratio compared with large particles and (b) easy function of organic moieties for improving the affinity of a compound. It was found that not all of the research activities that have been carried have been successful. Yet approaches are being continuously found to remove heavy metals on a large scale. Removal of heavy metal would be possible only if the community as a whole makes effort to develop absorption mechanisms and investigates more and more materials that can be used as next-generation absorbents. Several sections of society are highly concerned about the contamination of water by heavy metals, which leads to health issues, and are continuously working to find appropriate methods for heavy metal removal and thus eliminate associated health-related problems.

References

1. Elimelech, M., and Phillip, W. A. (2011). The future of seawater desalination: Energy, technology, and the environment. *Science*, **333**, pp. 712–717.
2. Zhang, S., Li, Y., Zhang, T., and Peng, Y. (2015). An integrated environmental decision support system for water pollution control based on TMDL–A case study in the Beiyun River watershed, *J. Environ. Manag.*, **156**, pp. 31–40.
3. Mohee, R., Mauthoor, S., Bundhoo, Z. M., Somaroo, G., Soobhany, N., and Gunasee, S. (2015). Current status of solid waste management in small island developing states: A review, *Waste Manag.*, **43**, pp. 539–549.
4. Ming, T., Liu, W., and Caillol, S. (2014). Fighting global warming by climate engineering: Is the Earth radiation management and the solar radiation management any option for fighting climate change?, Renew. *Sustain. Energy Rev.*, **31**, pp. 792–834.
5. Kurniawan, T. A., Sillanpää, M. E., and Sillanpää, M. (2012). Nanoadsorbents for remediation of aquatic environment: Local and practical solutions for global water pollution problems, *Crit. Rev. Environ. Sci. Technol.*, **42**, pp. 1233–1295.
6. Waghmare, S. S., and Arfin, T. (2015). Defluoridation by adsorption with chitin-chitosan-alginate–polymers–cellulose–resins–algae and fungi-a review, *Int. Res. J. Eng. Tech.*, **2**, pp. 1179–1197.
7. Arfin, T., and Mohammad, F. (2016). Chemistry and structural aspects of chitosan towards biomedical applications. In: *Natural Polymers: Derivatives, Blends and Composites*, vol. 1, eds., Ikram, S., and Ahmed, S., Chapter 13, (Nova Publishers, New York) pp. 265–280.
8. Mohammad, F., Arfin, T., and Al-Lohedan, H. A. (2017). Enhanced biological activity and biosorption performance of trimethyl chitosan-loaded cerium oxide particles, *J. Ind. Eng. Chem.*, **45**, pp. 33–43.
9. Arfin, T. (2017). Chitosan and its derivatives: overlook of commercial application in diverse field. In: *Chitosan: Derivatives, Composites and Applications*, eds., Ahmed, S., and Ikram S., Chapter 5, (Scrivener Publishing LLC) pp. 115–150.
10. Hernández-Morales, V., Nava, R., Acosta-Silva, Y. J., Macías-Sánchez, S. A., Pérez-Bueno, J. J., and Pawelec, B. (2012). Adsorption of lead (II) on SBA-15 mesoporous molecular sieve functionalized with $-NH_2$ groups, *Micropor. Mesopor. Mat.*, **160**, pp. 133–142.

11. Akpor, O. B., Ohiobor, G. O., and Olaolu, T. D. (2014). Heavy metal pollutants in wastewater effluents: Sources, effects and remediation, *Adv. Biosci. Bioeng.*, **2**, pp. 37–43.

12. Ngah, W. S. W., Ab Ghani, S., and Kamari, A. (2005). Adsorption behaviour of Fe(II) and Fe(III) ions in aqueous solution on chitosan and cross-linked chitosan beads, *Bioresour. Technol.*, **96**, pp. 443–450.

13. Aksu, Z., Calik, A., Dursun, A. Y., and Demircan, Z. (1999). Biosorption of iron (III)–cyanide complex anions to Rhizopus arrhizus: Application of adsorption isotherms, *Process Biochem.*, **34**, pp. 483–491.

14. Burke, A., Yilmaz, E., Hasirci, N., and Yilmaz, O. (2002). Iron(III) ion removal from solution through adsorption on chitosan, *J. Appl. Polym. Sci.*, **84**, pp. 1185–1192.

15. Wang, M., Xu, L., Zhai, M., Peng, J., Li, J., and Wei, G. (2008). γ-ray radiation-induced synthesis and Fe(III) ion adsorption of carboxymethylated chitosan hydrogels, *Carbohydr. Polym.*, **74**, pp. 498–503.

16. Nuhoglu, Y., and Oguz, E. (2003). Removal of copper(II) from aqueous solutions by biosorption on the cone biomass of Thuja orientalis, *Process Biochem.*, **38**, pp. 1627–1631.

17. Sun, S., and Wang, A. (2006). Adsorption kinetics of Cu(II) ions using N,O-carboxymethyl-chitosan, *J. Hazard. Mater.*, **131**, pp. 103–111.

18. Leyva-Ramos, R., Rangel-Mendez, J. R., Mendoza-Barron, J., Fuentes-Rubio, L., and Guerrero-Coronado, R. M. (1997). Adsorption of cadmium (II) from aqueous solution onto activated carbon, *Water Sci. Technol.*, **35**, pp. 205–211.

19. Salah, T. A., Mohammad, A. M., Hassan, M. A., and El-Anadouli, B. E. (2014). Development of nano-hydroxyapatite/chitosan composite for cadmium ions removal in wastewater treatment, *J. Taiwan Inst. Chem. Eng.*, **45**, pp. 1571–1577.

20. Wang, J. W., and Kuo, Y. M. (2008). Preparation and adsorption properties of chitosan–poly (acrylic acid) nanoparticles for the removal of nickel ions, *J. Appl. Polym. Sci.*, **107**, pp. 2333–2342.

21. Karthikeyan, T., Rajgopal, S., and Miranda, L. R. (2005). Chromium (VI) adsorption from aqueous solution by Hevea Brasilinesis sawdust activated carbon, *J. Hazard. Mater.*, **124**, pp. 192–199.

22. Gao, H., Liu, Y., Zeng, G., Xu, W., Li, T., and Xia, W. (2008). Characterization of Cr(VI) removal from aqueous solutions by a surplus agricultural waste—rice straw, *J. Hazard. Mater.*, **150**, pp. 446–452.

23. Sankararamakrishnan, N., Dixit, A., Iyengar, L., and Sanghi, R. (2006). Removal of hexavalent chromium using a novel cross linked xanthated chitosan, *Bioresour. Technol.*, **97**, pp. 2377–2382.

24. Kousalya, G. N., Gandhi, M. R., and Meenakshi, S. (2010). Removal of toxic Cr(VI) ions from aqueous solution using nano-hydroxyapatite-based chitin and chitosan hybrid composites, *Adsorpt. Sci. Technol.*, **28**, pp. 49–64.

25. Ghassabzadeh, H., Torab-Mostaedi, M., Mohaddespour, A., Maragheh, M. G., Ahmadi, S. J., and Zaheri, P. (2010). Characterizations of Co(II) and Pb(II) removal process from aqueous solutions using expanded perlite, *Desalination*, **261**, pp. 73–79.

26. Manohar, D. M., Noeline, B. F., and Anirudhan, T. S. (2006). Adsorption performance of Al-pillared bentonite clay for the removal of cobalt(II) from aqueous phase, *Appl. Clay Sci.*, **31**, pp. 194–206.

27. Chang, Y. C., Chang, S. W., and Chen, D. H. (2006). Magnetic chitosan nanoparticles: Studies on chitosan binding and adsorption of Co(II) ions, *React. Funct. Polym.*, **66**, pp. 335–341.

28. Bhattacharya, P., Welch, A. H., Stollenwerk, K. G., McLaughlin, M. J., Bundschuh, J., and Panaullah, G. (2007). Arsenic in the environment: Biology and chemistry, *Sci. Total Environ.*, **379**, pp. 109–120.

29. dos Santos, H. H., Demarchi, C. A., Rodrigues, C. A., Greneche, J. M., Nedelko, N., and Ślawska-Waniewska, A. (2011). Adsorption of As(III) on chitosan-Fe-crosslinked complex (Ch-Fe), *Chemosphere*, **82**, pp. 278–283.

30. Clarkson, T. W. (1993). Mercury: Major issues in environmental health, *Environ. Health Perspect.*, **100**, pp. 31–33.

31. Miretzky, P., and Cirelli, A. F. (2009). Hg(II) removal from water by chitosan and chitosan derivatives: A review, *J. Hazard. Mater.*, **167**, pp. 10–23.

32. Zhou, L., Liu, Z., Liu, J., and Huang, Q. (2010). Adsorption of Hg(II) from aqueous solution by ethylenediamine-modified magnetic crosslinking chitosan microspheres, *Desalination*, **258**, pp. 41–47.

33. Amarasinghe, B. M. W. P. K., and Williams, R. A. (2007). Tea waste as a low cost adsorbent for the removal of Cu and Pb from wastewater, *Chem. Eng. J.*, **132**, pp. 299–309.

34. Chen, H., Zhao, J., Dai, G., Wu, J., and Yan, H. (2010). Adsorption characteristics of Pb(II) from aqueous solution onto a natural biosorbent, fallen Cinnamomum camphora leaves, *Desalination*, **262**, pp. 174–182.

35. Wang, S., Dong, Y., He, M., Chen, L., and Yu, X. (2009). Characterization of GMZ bentonite and its application in the adsorption of Pb(II) from aqueous solutions, *Appl. Clay Sci.*, **43**, pp. 164–171.

36. Schneegurt, M. A., Jain, J. C., Menicucci, J. A., Brown, S. A., Kemner, K. M., Garofalo, D. F., Qualick, M. R., Neal, C. R., and Kulpa, C. F. (2001). Biomass byproducts for the remediation of wastewaters contaminated with toxic metals, *Environ. Sci. Technol.*, **35**, pp. 3786–3791.

37. Tan, Y., Chen, M., and Hao, Y. (2012). High efficient removal of Pb(II) by amino-functionalized Fe_3O_4 magnetic nano-particles, *Chem. Eng. J.*, **191**, pp. 104–111.

38. Yuan, L., and Liu, Y. (2013). Removal of Pb(II) and Zn(II) from aqueous solution by ceramisite prepared by sintering bentonite, iron powder and activated carbon, *Chem. Eng. J.*, **215–216**, pp. 432–439.

39. Liu, B., Lv, X., Meng, X., Yu, G., and Wang, D. (2013). Removal of Pb(II) from aqueous solution using dithiocarbamate modified chitosan beads with Pb(II) as imprinted ions, *Chem. Eng. J.*, **220,** pp. 412–419.

40. Kyzas, G. Z., Siafaka, P. I., Pavlidou, E. G., Chrissafis, K. J., and Bikiaris, D. N. (2015). Synthesis and adsorption application of succinyl-grafted chitosan for the simultaneous removal of zinc and cationic dye from binary hazardous mixtures, *Chem. Eng. J.*, **259**, pp. 438–448.

41. Tolaymat, T. M., El Badawy, A. M., Genaidy, A., Scheckel, K. G., Luxton, T. P., and Suidan, M. (2010). An evidence-based environmental perspective of manufactured silver nanoparticle in syntheses and applications: A systematic review and critical appraisal of peer-reviewed scientific papers, *Sci. Total Environ.*, **408**, pp. 999–1006.

42. Chen, X., and Schluesener, Nanosilver, H. J. (2008). A nanoproduct in medical application, *Toxicol. Lett.*, **176**, pp. 1–12.

43. Eckelman, M. J., and Graedel, T. E. (2007). Silver emissions and their environmental impacts: A multilevel assessment, *Environ. Sci. Technol.*, **41**, pp. 6283–6289.

44. Qi, X., Poernomo, G., Wang, K., Chen, Y., Chan-Park, M. B., Xu, R., Chang, M. W. (2011). Covalent immobilization of nisin on multi-walled carbon nanotubes: Superior antimicrobial and anti-biofilm properties, *Nanoscale*, **3**, pp. 1874–1880.

45. Hosoba, M., Oshita, K., Katarina, R. K., Takayanagi, T., Oshima, M., Motomizu, S. (2009). Synthesis of novel chitosan resin possessing histidine moiety and its application to the determination of trace silver by ICPAES coupled with triplet automated-pretreatment system, *Anal. Chim. Acta*, **639**, pp. 51–56.

46. Song, X., Li, C., Xu, R., and Wang, K. (2012). Molecular-ion-imprinted chitosan hydrogels for the selective adsorption of silver (I) in aqueous solution, *Ind. Eng. Chem.*, **51**, pp. 11261–11265.

47. Rajapaksha, A. U., Vithanage, M., Oze, C., Bandara, W. M. A. T., and Weerasooriya, R. (2012). Nickel and manganese release in serpentine soil from the Ussangoda Ultramafic complex, Sri Lanka, *Geoderma*, **189**, pp. 1–9.

48. Röllin, H. B., and Nogueira, C. M. C. A. (2011). Manganese: Environmental pollution and health effects, *Encycl. Environ. Health*, pp. 617–629.

49. Takeda, A. (2003). Manganese action in brain function, *Brain Res. Brain Res. Rev.*, **41**, pp. 79–87.

50. Al-Wakeel, K. Z., El Monem, H. A., and Khalil, M. M. (2015). Removal of divalent manganese from aqueous solution using glycine modified chitosan resin, *J. Environ. Chem. Eng.*, **3**, pp. 179–186.

Chapter 8

Bacterial Cellulose and Its Applications

Thazin Han,[a*] **Nitar New,**[b] **and Phyu Phyu Win**[a]

[a]*Department of Research and Innovation, Ministry of Education, Myanmar*
[b]*Ecological Laboratory, Advancing Life and Regenerating Motherland, Yangon, Myanmar*

thazinhann@gmail.com

Bacterial cellulose is a type of biopolymer, produced from several species of *Acetobacter* because of its unique properties. The popular strain *Acetobacter xylinum* is a key player for the production of bacterial cellulose. The cellulose derived from bacterial source is shortly called BC, the common name. Bacterial cellulose can be produced using various types of carbon-containing substrates such as honey, coconut water, coconut milk, palm water, papaya, pineapple, tomato, and some industrial wastes. The characteristics and productivity of bacterial cellulose mainly depend on the bacteria strain, carbon source, nitrogen source, pH, and temperature of culture media. Bacterial cellulose has numerous applications in industrial sectors as it is easy to produce

Handbook of Biopolymers: Advances and Multifaceted Applications
Edited by Shakeel Ahmed, Suvardhan Kanchi, and Gopalakrishnan Kumar

ISBN 978-981-4800-17-4 (Hardcover), 978-0-429-02475-7 (eBook)
www.panstanford.com

from low-cost culture and because of its qualified abilities, ease of decomposition by microorganisms in the environment, and renewability. Therefore, it has been also applied in many industrial processes such as the food industry, medical materials industry, paper industry, textile industry, electrical applications, and composites industry.

8.1 Introduction

Cellulose is the most abundant natural biopolymer in the nature, produced from plants, algae, some species of bacteria, and some animals [1]. Many species of Gram-negative bacteria, such as those in the genera *Agrobacterium, Aerobacter, Komagataeibacter rhaeticus, Enterobacter, Escherichia, Gluconacetobacter, Azotobacter, Rhizobium, Sarcina,* and *Salmonella* and several species of *Cyanobacteria* are likely to produce extracellular cellulose [2–4]. Among Gram-negative organisms, *Acetobacter xylinum* is the most efficient bacterium to be an effective cellulose-producing bacterium and is widely used in industrial productions [5]. It is a simple Gram-negative obligate aerobe. Cellulose synthesis in *Acetobacter xylinum* was discovered by Brown over 100 years ago. It is known that a gelatinous membrane was found in the process of producing vinegar. The gel-like thin mat, which has produced from *Acetobacter xylinum* bacterium, was formed in the liquid medium.

Bacterial cellulose is a polysaccharide synthesized by *Acetobacter xylinum*; it grows on the air–liquid surface of culture medium not only in a static culture but also in shaking condition and stirring bioreactor [6]. The genus *Acetobacter* is a group of bacteria that have the ability to oxidize sugars, alcohols, and acetic acid as a major end product [7]. Cellulose-producing *Acetobacter* are commonly found in natural sources such as vegetables, flowers, sugar cane, nuts, and rotten fruits [8]. The strains of *Acetobacter xylinum* are capable of producing cellulose in varying amounts and growing medium on the wide variety of substrates such as glucose, sucrose, fructose, invert sugar, ethanol, and glycerol. *Acetobacter xylinum* has been shown to be distinctly acid-tolerant organism at pH as low as 3.5 and reported pH 4.0 and 5.0 to be ideal for the development of cellulose. Bacterial cellulose can also be grown in many culture mediums of waste [10]. The cheapest source of media, coconut water is one of the culture media usually used to produce bacterial cellulose by *Acetobacter xylinum*. A single *Acetobacter xylinum* is capable of polymerizing 200,000 glucose molecules per second into β-1,4-glucan chains, which are then excreted into the surrounding medium forming ribbon-like bundles of microfibrils [11]. Bacterial cellulose possesses very high crystallinity, high tensile strength, extremely insolubility in most of the solvents, mold ability, and high degree of polymerization [12, 13]. The thickness of cellulose fibrils is commonly 0.1–10 μm, one hundred times thinner than cellulose fibrils obtained from the real plant [14].

Water-absorbing ability of bacterial cellulose was outstanding compared with other organics. Normally, bacterial cellulose could contain water 60–700 times heavier than its own dry weight. Because of the superfine structure of bacterial cellulose and its hydrophilic group, many "pores" were formed in the cellulose membranes to hold water. It was about 300 times more than wood cellulose [15]. Bacterial cellulose is far stronger than plant cellulose [16]. The macroscopic morphology of cellulose depends on the condition of the culture, which can easily be tailored for physicochemical properties [5]. The culture conditions for bacterial cellulose production such as type of strain, temperature of growth condition, carbon source, nitrogen source, pH, and exposition time are also affected the physical properties of the bacterial cellulose production. The porosity and thickness of resulting bacterial cellulose can be varied according to the purification conditions of bacterial cellulose [17].

8.2 Bacterial Cellulose Production

Bacterial cellulose is industrially produced from *Gluconobacter xylinum* (*Acetobacter xylinum*) and some *Acetobacter* spp. *Acetobacter* bacteria have been commonly found in symbiotic relationships with many different plants and fruits such as sugarcane, coffee plants, rotten fruits, and consequently isolated. Table 8.1 shows that the type of bacterial strain and yield of bacterial cellulose [18]. *Acetobacter xylinum,* an obligate aerobe, to grow in the higher oxygen tension at the surface of the medium, produces two forms of cellulose: (i) cellulose I, the ribbon-like polymer, and (ii) cellulose II, the thermodynamically more stable amorphous polymer [18–20]. Figure 8.1 shows cellulose microfibrils I and II produced by *Acetobacter xylinum* [21].

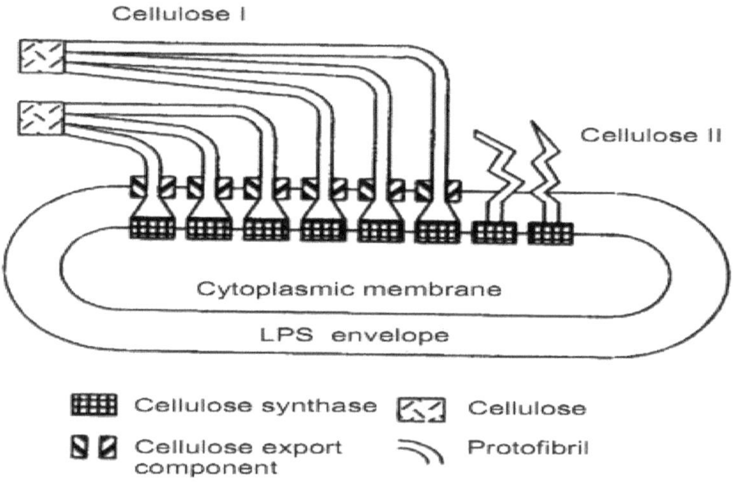

Figure 8.1 Assembly of cellulose microfibrils by *Acetobacter xylinum* [16, 21].

The bacterial cellulose is crystallized into fibers and its fibers resemble in width and structure the average fibrils from many plants and algae. The fibers are formed in the membrane by cellulase and synthase and consequently secreted from a row of 50–80 pore-like synthetic sites along the longitudinal axis of the cell. Figure 8.2 describes atomic force microscopy (AFM) surface monograph of dried bacterial cellulose from coconut medium, which indicates that there is no crack and hole on the surface of it. The formation of this floating cellulose matrix is thought to allow *Acetobacter xylinum* [22]. This bacterium was cultured by certain amount of nutrient medium and suitable condition of culture medium [23]. It was carried out under the optimized conditions of carbon source, nitrogen source, mineral source, temperature, pH, shaking, and fermenter design. Depending on the bacteria strain, carbon, nitrogen, mineral sources, temperature, pH, and duration of culture, bacterial cellulose productivity varies [7, 24, 25].

Glucose, sucrose, and other carbohydrates such as fructose, maltose, xylose, starch, and glycerol have been used as carbon sources for the production of bacterial cellulose. The effect of initial glucose concentration on cellulose production is also important since the formation of gluconic acid as a byproduct in the medium that tends to decrease the production of cellulose. The 28% of total consumption of glucose was found to be used for cellulose production [26]. Fructose was found to be the best carbon source at 2% concentration for the production of bacterial cellulose [27]. Mannitol has been reported to be the best carbon source for the new bacteria strain *Acetobacter* sp. DR-1 as it supported the maximum cellulose production [7].

Bacterial cellulose was mainly produced from *Acetobacter xylinum* using fruits wastes as substrate of culture such as coconut water, papaya juice, and muskmelon juice (wastage), where papaya juice without additional sugar proved to be an efficient substrate for the production of bacterial cellulose compared with substrates that have additional sugar 4.52 g/100 ml. The carbohydrate analysis of papaya juice, coconut water, and muskmelon juice showed the presence of 2.2%, 1.8%, and 1.5% of total sugar, respectively. The highest yield of bacterial cellulose was obtained from the media with coconut water, papaya juice, or muskmelon without additional sugar, compared with these substrates with additional sugar. Hence, papaya juice used as natural carbon source, i.e., without additional sugar, gave the highest yield of bacterial cellulose followed by coconut water, and muskmelon juice gave the lowest cellulose yield. Cellulose production increases with the use of particular carbon sources in some media, but not in others, and the yield is greatly affected by the selection of the media. The static condition of culture is suitable for bacterial cellulose production [28]. Bacterial cellulose production also depends on the amount of carbon and nitrogen, which provide nutrients for the growth of *Acetobacter xylinum*. Comparing the use of glucose and sucrose in the medium, it was found that the use of glucose in the medium produces thicker bacterial cellulose than other carbon sources [29]. However, when bacterial cellulose was produced from beet molasses using *Gluconacetobacter xylinus* (ATCC 10245), the yield of the bacterial cellulose produced was high compared with glucose as the sole carbon source [30]. Therefore, the production of bacterial cellulose mainly depends on the carbon source and the species of bacteria.

Another report has concluded that bacterial cellulose was produced from *Gluconacetobacter xylinus* (PTCC, 1734) in sucrose and date syrup solutions (Bx. 10%) under static conditions at 28°C. The maximum yields of dried bacterial cellulose (BC) were 4.35 and 1.69 g/100 ml for date syrup and sucrose medium after a 14-day fermentation period [31]. An edible film of bacterial cellulose was produced using *Acetobacter xylinum* in the medium of coconut water and *Arenga pinnata* juice. A different ratio of *Arenga pinnata* juice and coconut water in the culture medium has also been used. In another study, *Acetobacter xylinum* (TISTR 975) was cultured in a coconut water medium with added $MgSO_4$ and NH_4HPO_4. The optimum pH and the amount of added sugar for the production of cellulose were found to be 4.75 and 5.1% w/v, respectively, and the production of dried cellulose was measured after 8 days of fermentation; the depth of the liquid medium and the surface area varied. The depth of the medium for the highest wet cellulose production was found to be 2.5 cm. The increment of the medium volume by the depth of the liquid medium was less effective for the production of cellulose than the increasing surface area. The highest amount of dried cellulose produced was 5.94 g/L from the fermentation conditions of 2.5 cm depth and liquid/air interface area of 359.12 cm^2 [32]. The production of bacterial cellulose from *Gluconacetobacter persimmonis* GH-2 was able to use various carbohydrates (2% w/v in HS medium) from various natural carbon sources for growth and cellulose production. All natural carbon sources were able to support growth and cellulose production. When molasses, watermelon, orange juice, and muskmelon were provided as carbon and nutrient sources, the organism was able to produce 5.75, 5.98, 6.18, and 8.08 g/ L of dried bacterial cellulose. The maximum bacterial cellulose yield was given by muskmelon [33].

The nitrogen sources that can be used as a bacterial cellulose culture medium are peptone, casein hydrolyzate, urea, ammonium sulfate, and sodium nitrate. The addition of extra nitrogen favors the biomass production but diminishes cellulose production [29]. The effect of various nitrogen sources on the production of bacterial cellulose has been reported; casein hydrolyzate gave cellulose yield of 4.8 g/L of medium [33]. Peptone supported the maximum cellulose yield compared with the rest of the nitrogen sources [7].

Table 8.1 Cellulose-producing bacteria and yields of cellulose [18]

Bacteria	Carbon source	Supplement	Culture time (h)	Yield (g L^{-2})
A. xylinum BRC 5	Glucose	Ethanol+oxygen	50	15.30
G. hansenil	Glucose	Oxygen	48	1.72
G. hansenil	Glucose	Ethanol	72	2.50
Acetobacter sp. V6	Glucose	Ethanol	192	4.16
Acetobacter sp. V9	Glucose	Ethanol	192	15.20
A. xylinum BPR2001	Molasses	None	72	7.82
A. xylinum BPR2001	Fructose	Agar oxygen	72	14.10
A. xylinum BPR2001	Fructose	Agar	56	12.00
A. xylinum ssp. sucrofermentans BPR2001	Fructose	Oxygen	52	10.40
A. xylinum ssp. sucrofermentans BPR2001	Fructose	Agar oxygen	44	8.70
A. xylinus E25	Fructose	No	168	3.50
G. xylinus K3	Mannitol	Green tea	168	3.34
G. xylinus IFO 13773	Glucose	Lignosulfonate	168	10.10
A. xylinum NUST4.1	Glucose	Sodium alginate	120	6.00
G. xylinus IFO 13773	Molasses	No	168	5.76
Gluconacetobacter sp. RKY5	Glycerol	No	144	5.63
Co-culture of *Gluconacetobacter* sp. st-60–12 and *Lactobacillus mali* JCM1116	Sucrose	No	72	42.0

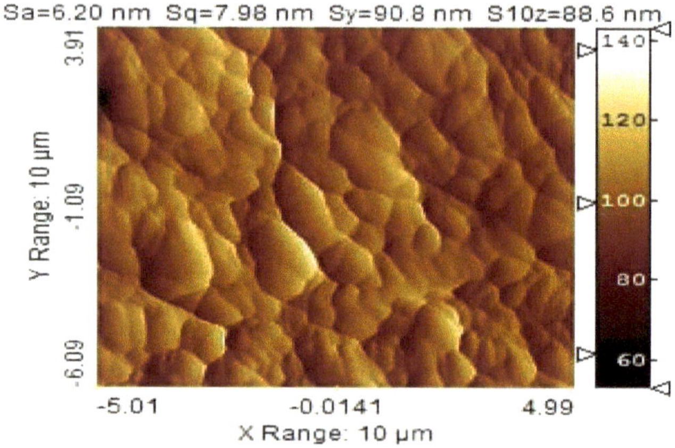

Figure 8.2 AFM surface morphology of dried bacterial cellulose from coconut medium. Adapted from the author's unpublished research.

Temperature is a crucial parameter that affects both growth and cellulose production. The best temperature for the maximum yield of cellulose production was observed to be 20–25, 28, and 30°C [7, 34–37]. The variation in temperature caused changes of cellulose degree of polymerization and water-binding capacity, whereas, the bacterial cellulose synthesized at 30°C had a lower DP (~10,000) and a higher water-binding capacity (164%) compared with that produced at 25 and 35°C [35]. However, there was no significant difference in cellulose production at 25°C, but cellulose production decreased above 35°C [38]. The optimum temperature was obtained at 30°C and the lowest yield was found to be at 45°C [39].

It was reported that the optimum pH of the culture medium for bacterial cellulose production was in the range of 3.5 to 7 depending on the type of bacteria. The cellulose production with bacteria was observed to be the maximum at pH 6.5–7 and the lowest yield was found to be at pH 4 [38]. It was found that the yield of bacterial cellulose from *Acetobacter xylinum* decreased below pH 4 [23]. With the use of *Acetobacter pasteurianus* HBB6 and *Acetobacter lovaniensis* HBB5 strains, the highest yield was found at pH 6.5 and the lowest yield at pH 3.5 and 4.5 in Herstin–Schramm medium or beet molasses, respectively. The bacterial cellulose yield was not observed at pH 2.5 [35]. The optimal pH range for cellulose production by *Acetobacter xylinum* is 4 to 7 [39–41]. The pH decreases during fermentative production because of the accumulation of gluconic, acetic, or lactic acids in the culture broth [42].

The fermenter design is also important for the productivity of bacterial cellulose; it withstands vigorous mechanical agitation of the rapidly growing *Acetobacter xylinum* (obligate aerobe) culture and also prevents the cellulose fibril and fibrillar matrix from mechanical disruption. The desirable modifications in the fibrillar and/or macroscopic nature of the cellulosic product could be achieved by varying fermenter design factors such as the shape of vessels and agitating impellers. Use of some modified reactors such as a rotating-disk reactor, rotary biofilm contactor, bioreactors equipped with a spin filter, and reactor with silicone membrane has also been reported. Shake flask culture supported maximum cellulose yield as it increased the oxygen content of the flask, and literature supports that aeration is needed for the maximum cellulose production. An economical mass production system of bacterial cellulose on agitated culture was constructed [36, 37]. The shake flask culture was better than the conventional method for the production of bacterial cellulose [37–42].

The thickness of bacteria cellulose varied; it depends on the amount of culture medium and duration of fermentation [43]. The duration of bacteria cellulose fermentation can vary depending on the production of the desire material. The time profile of the maximum cellulose production by the strain *Acetobacter* sp. DR-1 in sugarcane juice media under optimized conditions is the 10th day; thereafter, there is no increase in cellulose yield [7]. The greatest increase in the weight of bacterial cellulose takes place after 7–8 days of breeding *Acetobacter xylinum* at a temperature of 30°C, using a Herstin–Schramm nutrient medium containing: glucose 2 w/v%, yeast extract 0.5 w/v%, bacto-peptone 0.5 w/v%, citric acid 0.115 w/v%, Na_2HPO_4 0.27 w/v%, $MgSO_4 \cdot 7H_2O$ 0.05 w/v%, and ethanol 1 w/v%, added after the sterilization of the base [43]. *Acetobacter xylinum* TISTR976 was cultured in coconut juice medium to produce high-yield bacterial cellulose at optimized conditions [44]. Some reports showed that bacterial cellulose has been produced from *Acetobacter xylinum*, grown on the broth of sugar cane, fruit juice, industrial waste, or tea. It was indicated that sugarcane juice supported the highest yield of dried cellulose compared with the synthetic media used. A yield of 1.68 g/L of dried cellulose was obtained from sugarcane-based media, while it was 1.36 g/L from semi synthetic media [7]. For the industrial production of bacteria cellulose for nata de coco, coconut water is utilized as a substrate; the favorable duration of fermentation is 14–21 days for 1.2 liter tray containers, and the yield of wet bacterial cellulose is 1 kg [44].

Aseptic environment is also very important for the synthesis of bacterial cellulose for food and medical applications [45].

8.3 Application of Bacterial Cellulose

8.3.1 Applications of Bacterial Cellulose in Food Sector

Bacterial cellulose in food sector is popularly known as *nata de coco*. It was first developed in the Philippines as an unconventional product based on coconut water in 1949. The Philippine dessert nata de coco has been manufactured as a cottage industry. The import of nata from the Philippines into Japan has had a major impact on the global outlook for the expansion of microbial cellulose production [42]. Over the years, it has become popular in other countries such as Japan, Indonesia, Korea, Thailand, and the United States [46]. Figure 8.3 shows that the actual static culture of nata de coco at a production site in Myanmar. Figure 8.4 describes that bacterial cellulose as nata de coco in the laboratory culture conditions and small-scale production in optimum conditions.

The liquid is trapped in the cellulosic matrix, which is important for foods, and it is tasty. The thickness of nata influences the water-holding capacity, which in turn has a bearing on the physical textural and organoleptic properties. Nata de coco has distinct textural properties, such as a firm chewy, soft and smooth surface, which is preferred. When nata de coco is stressed, it releases its water and deforms without fracture. It can absorb water again when it is dipped in water. As nata de coco is mainly composed of cellulose, it has unique properties, including high mechanical strength, crystallinity, and water retention ability in the drink and cream. Furthermore, it has high purity and is free of hemicelluloses and lignin compound [22].

Figure 8.3 The actual static culture of nata de coco production in the cottage industry. Adapted from the author's unpublished research.

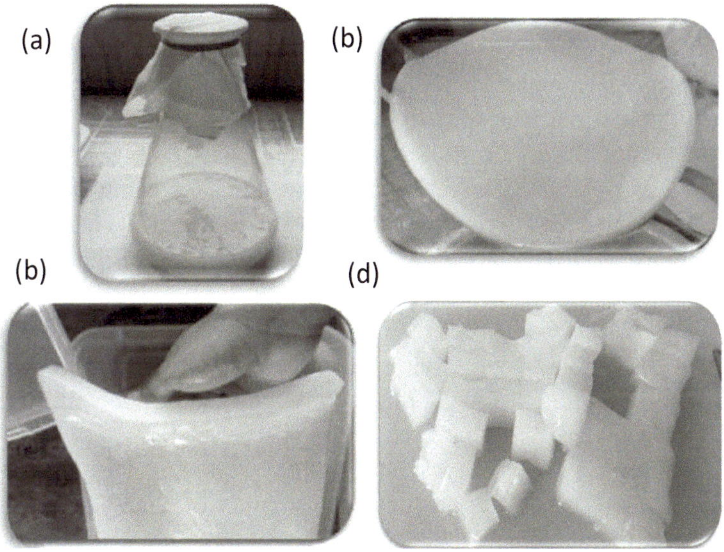

Figure 8.4 Bacterial cellulose (BC) produced from coconut water as substrate and with static condition at 28–30°C. (a) Lab-scale BC culture in conical flask. (b) Large-scale BC culture in large container. (c) Industrial culture for nata de coco production. (d) Nata de coco cubes for foods. Adapted from the author's unpublished research.

Initially, nata de coco was produced from expensive culture media, containing glucose as carbon source and other nutrient sources resulting in very high production costs, which limits the use of the material to very high-value applications. The fermentation process using the coconut water medium was implicated to produce nata de coco by *Acetobacter xylinum* in around 15–20 days in a static condition [47, 48].

Apart from coconut water, other unconventional media such as corn steep liquor, juice of fruit skin, and fruit wastes are used for the production of nata [49, 50]. Sucrose concentration of 10% and ammonium sulfate concentration of 0.5% at pH 4.0 were found to be the optimum conditions for nata de coco and could help achieve the maximum thickness of nata [22]. The use of cheaper carbon and nutrient sources is an interesting strategy to overcome the limitation of production cost and therefore to increase the competitiveness of this unique material. Nata can be produced not only from coconut water but also from most fruit juices and some fruit wastes. Mostly fruits are sold and consumed as raw food, but most of the damaged fruits and those of non-standard size are shelved, though some are processed to make jams, pastes, and sauces. When fruits cannot be shipped because of their poor quality caused by bad weather and some other natural disasters, it leads to low prices and fruit wastage. The majority of waste ends up being discarded. However, fruits have abundant sugars such as glucose and fructose, which could be bio-converted into bacterial cellulose [9]. Nata de coco is used in a variety of food formulations such as an additive, thickener, disperser, emulsifier, dietary fiber, and edible preservative. The gel itself is too hard to bite, but it becomes edible through processing either with sugar syrup or with drinks and fruit salad [51].

Nata de coco has specific abilities such as lack of flavor interactions, foam stabilization, and stability over a wide range of pH, temperature, and freeze-thaw conditions. The potential uses of nata de coco include pourable and spoonable dressings, sauces and gravies, icings, sour cream, cultured dairy products, aerated desserts, and frozen dairy products [52]. Nata de coco introduced as a diet drink through *A. xylinum* showed that it has a plasma cholesterol-lowering effect [53].

Due to tensile strength, low oxygen transmission (barrier property) rate, and its hydrophilic nature, the processed bacterial cellulose membrane appears to be of great relevance for its application as a packaging material in food packaging [54].

8.3.2 Applications of Bacterial Cellulose in Biomedical Sector

Bacterial cellulose does not involve toxic and any kind of hazardous materials for human beings, which is excellent and suitable for safe environments such as in medical and cosmetic applications [18]. Bacterial cellulose is receiving great attention and is being widely investigated for drug delivery, as a new type of scaffold materials due to its fine fiber network, biocompatibility, high water-holding capacity, high tensile strength, high crystallinity, high degree of polymerization, high purity, elasticity, durability, and non-toxic and nonallergic nature [55–60]. Bacterial cellulose has a wide range of medical applications, which include temporary artificial skin for the therapy of burns, ulcers, and dental implants [61] audio headphone diaphragm and wound healing [62–65]. Recently, bacterial cellulose has been used in many special applications such as a scaffold for tissue engineering of cartilages [66–70] and blood vessels [71, 72], as well as for artificial skin for temporary covering of wounds [73], nerve regeneration [74], and vascular grafts [75, 76].

8.3.2.1 Drug delivery system

Bacterial cellulose itself has no antimicrobial activity to prevent wound infection. The properties of a wound-healing system can easily be oriented toward transdermal drug delivery as it prevents moisture from evaporating, avoids external contamination, and maintains intimate contact with the exposed, inflamed, or diseased area. This facilitates localized drug delivery to the target site [77–80]. Figure 8.5 shows the representation of processes ranging between sources of bacterial cellulose and drug delivery system.

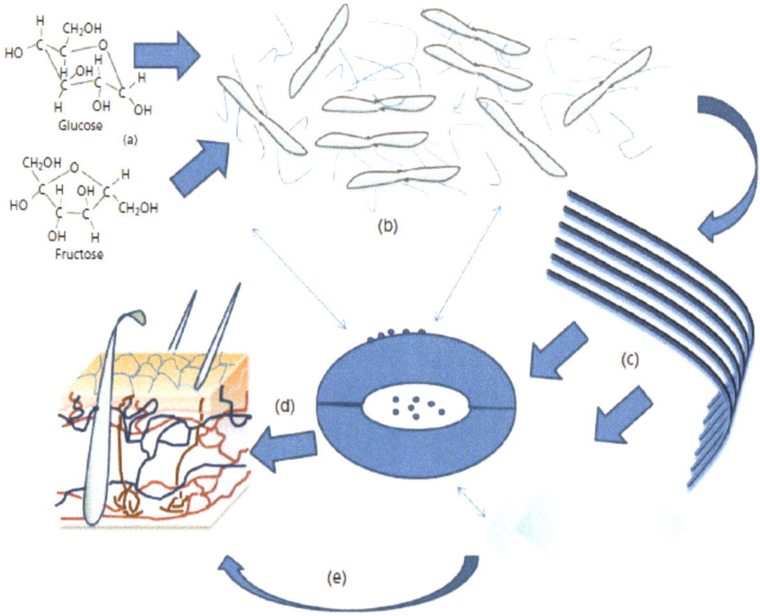

Figure 8.5 Representation of processes ranging between sources of bacterial cellulose and its potential applications. (a) The glucose and fructose molecules; representatives of carbon sources, which have been the main determinant in molecular properties of bacterial cellulose from *Gluconacetobacter xylinus*. (b) Biosynthesis of bacterial cellulose shown here as blue fibers (subfibers and microfibrils) originating from the replicating bacteria. (c) Bacterial cellulose microfibrils from *Gluconacetobacter xylinus* shown as a product of electrolyte-free system, which forms this shape that is non-chiral nematic phase. The bacterial cellulose fibers as protective membranes or drug delivery carriers. The latter is shown by entrapped and absorbed drug molecules in the longitudinal section of a single bacterial cellulose fiber. (d, e) The two paths show the prevalent use of bacterial cellulose products on skin/ topical surface (however, it is not to exclude the use of bacterial cellulose for designing oral and other drug delivery systems). Double-headed arrows explain that all the things are connected to a central point somehow, which focuses on the potential of bacterial cellulose as drug delivery system [69].

Both to deliver drug and to absorb exudates, transdermal delivery systems that can work bilaterally represent a very exciting opportunity for the application of bacterial cellulose membranes. The delivery activity is related to the diffusion potential of bacterial cellulose membranes. It has been investigated by loading tetracycline in electron beam-irradiated and -non-irradiated samples. The diffusion of drug enriched to solvent-only cells was found to be convective. The system comprised two poly(methyl-methacrylate) compartments that were separated by an operational membrane. The irradiated bacterial cellulose was confirmed by a decrease in the diffusion coefficient of irradiated samples; however, the non-irradiated bacterial cellulose allowed faster drug movement. It was proposed that a model for the adsorption of drugs onto bacterial cellulose membrane [81] and the drug delivery was assessed by placing the membrane on irregular surfaces and measuring drug deposition uniformity through the epidermis; the permeation rate for ibuprofen, a lipophilic drug, was found to be three times higher than that of lidocaine hydrochloride [82].

Bacterial cellulose could also be applied in conjunction with a conducting polymer, polyaniline. Since it possesses electro-conductivity and the potential to act as an electrically stimulated drug delivery device, the bacterial cellulose composite was constructed by the polymerization of aniline on a single side of the bacterial cellulose membrane and this composite acted as a supercapacitor [83]. This property could therefore be conducted with biocompatibility and biodegradability and improved the mechanical properties of bacterial cellulose, although the hydrophobicity of the conducting polymers needs to be modified for successful protein entrapment. Bacterial cellulose was constructed with multiwalled carbon nanotubes and electro-conductivity was increased. The iontophoresis (drug delivery by electric potential) can be applied to expand the application of bacterial cellulose [84].

To achieve antimicrobial activity for bacterial cellulose construction, silver nanoparticles have been impregnated into bacterial cellulose through the chemical reduction by immersing bacterial cellulose in the silver nitrate solution. Sodium borohydride was used to reduce the absorbed silver ion (Ag^+) inside bacterial cellulose and to increase absorbed metallic silver nanoparticles (Ag^0) inside bacterial cellulose. The antimicrobial activity of freeze-dried silver nanoparticle-impregnated bacterial cellulose prepared from the $NaBH_4$: $AgNO_3$ molar ratio of 100:1 against *Escherichia coli* and *Staphylococcus aureus* has been reported.

The freeze-dried silver nanoparticle-impregnated bacterial cellulose exhibited a strong antimicrobial activity against both *S. aureus* (Gram-positive bacteria) and *E. coli* (Gram-negative bacteria), which are general bacteria found on the contaminated wound [85]. It was also found that bacterial cellulose fibers with silver nanoparticles successfully produce up to 99.99% antibacterial activity against *E. coli* and *S. aureus* [86]. The use of silver nanocomposites with bacterial cellulose has been endorsed as a promising strategy in which antibacterial properties are required [87, 88].

Apart from silver nanoparticles, bacterial cellulose dry films have been investigated with benzalkonium chloride (an antimicrobial agent), after immersion in benzalkonium chloride, where the drug-loading capacity per unit surface area was 0.116 mg/cm^2, and drug effects lasted for at least 24 h against contaminated wound flora such as *S. aureus* and *Bacillus subtilis* [89]. A transdermal patch could be worked for enantiomeric release by constructing a gel reservoir and molecularly imprinted polymer (MIP) membrane. MIP membranes alone in comparison with the non-imprinting polymer (NIP) membrane and cellulose membranes were investigated for in vitro drug release to get the gel reservoir. It was subjected to in vitro skin permeation with the dorsal skin excised from male *Wistar* rats so that the control imparted on release could be identified. The *S*-enantiomer of propranolol (an antihypertensive drug) has been released from a composite layer of bacterial cellulose membrane with methacrylate and applied to transdermal application where the primary control of drug release came from the parent bacterial cellulose membrane. Herein, NIP was composed of a bacterial cellulose membrane with reactive pore filling by *S*-propranolol. This process was found to confer the superior character of enantioselective release to MIP over the NIP-cellulose membranes [90]. Furthermore, a cumulative release of 60% by chitosan and 48% by poloxamer was also reported.

The enhanced enantioselectivity of NIP composite membrane was attributed to the increase in the degree of ionization of the functional monomer, which increased the binding of the *S*-propranolol enantiomer at the imprint site. In addition, the type of gel used was found to affect the enantioselective transport of propranolol, with the more rheologically structured poloxamer gel formulation providing no selective release of *S*-propranolol. In vivo investigations of transdermal patch gel reservoirs composed of chitosan and poloxamer demonstrated that after 8 h, the C_{max} of *S*-propranolol (8.0 ± 1.0 ng/ml) was achieved from a reservoir of 1.5 mg of racemic propranolol [91]. These encouraging results indicated that bacterial cellulose, which acts as a base for molecular imprinting networks, is amendable to treatments such as reactive pores filling, which makes it suitable for chiral applications.

It was found that the two lines of cellulose with different charged moieties could be linked to carboxymethyl cellulose (CMC) using sodium (Na^+) and/or ammonium (NH_4). Compared with a chitosan hydrogel, the cellulose-based system was more favorable for buprenorphine with sodium; the bulkiness of the ion was the underlying reason, with sodium being the least bulky with corresponding higher mobility. The enhancement in permeation was achieved, but it was lower with sodium carboxymethyl cellulose compared with that with others, but CMCNa was the most effective for overall iontophoretic permeation [92]. Iontophoresis improved permeation by a factor of 14.27, compared with passive diffusion with strong electrophoretic drift of the buprenorphine molecule, which explains differences in the permeation and flux results. Passive diffusion was found to be less effective than iontophoresis, where electroporation was also employed. By using dorsal skin of female Balb\c-nu mice, the formation of transient aqueous pores was encouraging; the same mechanics were obtained from stratum corneum-stripped skin [93].

A series of bacterial cellulose hydrogels were made into membranes for the permeation of flurbiprofen, a lipophilic drug. The drug release with the concurrent use of transdermal enhancers and retarders such as limonene and phospholipids helped to achieve a drug release equivalent to that in the aqueous buffers. The membrane as a reservoir was lower with cellulose hydrogel. In vitro studies with various skin models harvested from female Balb\c-nu mice, such as de-lipidized skin, enhancer/retarder pretreated skin and stratum corneum stripped skin, were performed to characterize flurbiprofen absorption [94]. Bacterial cellulose has also been applied for its suitability in drug-eluting bone cements. The existing bone cements display limited drug release in terms of the temporal characteristics of release and have been limited by reduced mechanical strength. In this regard, bacterial cellulose was applied in traditional bone cement along with vancomycin hydrochloride and gentamicin sulfate demonstrating that the combination yield improved the sustained release compared with traditional cements. The increase in the duration of release could also be attributed to the increased levels of antibiotic concentration in the modified bone cement. Due to the presence of bacterial cellulose, the inferences were in addition to improved mechanical properties such as compression strength and fatigue [95].

Most of bacterial cellulose products have already been approved by the Food and Drug Administration (FDA) as a component of medical products used for various clinical indications (Table 8.2). The U.S. FDA's report clearly indicates that bacterial cellulose is the primary and sole component of some approved products for wound dressing, a fact that assists us in our effort to promote its use for potential fabrication of drug delivery systems. There are a number of patents of bacterial cellulose, which have been registered to protect the intellectual property of effective or potentially effective drug delivery systems that possess a wide range of applications [96], and they are also described in Table 8.3 [68].

Table 8.2 Products based on bacterial cellulose available in market [69]

Product	Type	Clinical indication/use	Company/agency
Basyc	Vessel implant (tubes)	CABG (Coronary artery bypass surgery)	Jenpolymer materials Ltd. & Co.
Bio Fill	Wound care system	Burns	Robin goad
Bioprocess	Artificial skin	Burns	Biofill Produtos Biotechnologicos
Cellulon	Binder	Medical applications Including non woven structures	CP Kelco

Product	Type	Clinical indication/use	Company/agency
Cellulon PX microfibrous cellulose	Suspending agent	Suspension of particles, encapsulated enzymes	CP Kelco
CelMat (C) MG & CelMat (R) MG	Protective dressings/ jackets	Protection for miners from potential burns	Government of Poland
Dermafill	Wound care dressing	Burns	Fibrocel
Gengiflex	Non-resorbable cellulose membrane	Periodontitis	Biofill Produtos Biotechnologicos
MTA protective tissue	Biocompatible implant	Injury and wound care	Xylos corporation
Securian	Tissue reinforcement matrix	Tendon repair	Xylos corporation

MTA, Mineral Trioxide Aggregate.

Table 8.3 Patents based on bacterial cellulose products for drug delivery [69]

Field of invention	Patent title	Patent number	Registration
Implantable device; soft tissue repair-drug delivery carriers	A method for producing implantable microbial cellulose materials for various medical applications	EP1795213 B1	European patent office (EPO)
Network meshed hydrogel, Drug delivery carrier, skin substitute	Novel network meshed hydrogel structure	TW M428771U1	Intellectual Property Office Taiwan (TIPO)

(*Continued*)

Table 8.3 (*Continued*)

Field of invention	Patent title	Patent number	Registration
Calcium alginate capsules embedded and prepared in situ; containing drugs, probiotics	Bacterial cellulose composite with capsules embedded therein and preparation thereof	US 2012308649A1	United states patent and trade mark office (USPTO)
Implantable bacterial cellulose; in-vivo application	Thermally modified microbial-derived cellulose for in-vivo implantation	EP1662976 A2 US20050042250 US8198261	USPTO, 2006 & EPO, 2005
Use of microbial (bacterial) cellulose in transdermal drug delivery	Microbial cellulose materials for use in transdermal drug delivery systems, method of manufacture and use	US 20060240084	USPTO, 2006
Medical implant; orthopedic	Medical device including bacterial cellulose reinforced by resorbable or non-resorbable materials	US 20110262521A1	USPTO, 2011
Wide range of applications, dependent on density gradient dictated by thickness; number of drugs can be delivered	Bacterial cellulose films and uses thereof	EP 2390344 A1 US20110286948	EPO, 2011 USPTO, 2011

Field of invention	Patent title	Patent number	Registration
Cellulose hydrogels, making and applications; implant and ocular devices, sustained release drug delivery systems	Cellulose-based hydrogels and methods of making thereof	US20130032059 A1	USPTO 2013

8.3.2.2 Bacterial cellulose scaffold for tissue engineering

Bacterial cellulose is an excellent candidate for tissue engineering scaffold as evidenced by many studies in which bacterial cellulose nanofibers with appropriate porosity were investigated for their application as biocompatible cell adhesion and proliferation supports [97, 98]. Figure 8.6 shows that the wet condition of pure bacterial cellulose scaffold. The application of scaffolds in the tissue engineering of cartilage is essential in order to support cell proliferation and maintain their differentiated function in addition to definition of the shape of the new growing tissue. Certain strain of *Acetobacter xylinum* was explored as a novel scaffold material due to its unusual material properties and degradability [99]. The evaluation of native and chemically modified bacterial cellulose materials (phosphorated and sulfonated) in the using of bovine chondrocytes was reported. This evaluation showed that unmodified bacterial cellulose supports chondrocyte proliferation at levels of approximately 50% of the collagen type II substrate while providing significant advantages in terms of mechanical properties. Compared to tissue culture plastic and calcium alginate, unmodified bacterial cellulose showed significantly higher levels of chondrocyte growth. Chemical sulfonation and phosphorylation of bacterial cellulose, performed to mimic the glucose amino glycans of native cartilage, did not enhance chondrocyte growth, while the porosity of the material did affect chondrocyte viability. Bacterial cellulose did not induce significant activation of pro-inflammatory cytokine production during in vitro macrophage screening. Hence, unmodified bacterial cellulose was further explored using human chondrocytes [100]. According to the transmission electron microscopy (TEM) analysis and RNA expression of the collagen II from human chondrocytes, the unmodified bacterial cellulose supported proliferation of chondrocytes. The growth of chondrocytes into the scaffold by TEM, the unmodified bacterial cellulose would be used as a scaffold for tissue engineering of cartilage [101].

Figure 8.6 The pure wet bacterial cellulose scaffold before drying. Adapted from the author's unpublished research.

8.3.2.3 Bacterial cellulose membrane as skin therapy

Bacterial cellulose has the properties of high mechanical strength in the wet state, substantial permeability for liquids and gases and low irritation of skin so that the gelatinous membrane of bacterial cellulose was usable as an artificial skin for temporary covering of wounds [102, 103]. There are several bacterial cellulose products in the market. The products of bacterial cellulose, Biofill® and Gengiflex® have been applied in surgery and dental implants and realities in the human health-care sector. Cases of second and third degree of burns, ulcers and others could be treated successfully with Biofill® as a temporary substitute for human skin. The advantages of Biofill® in more than 300 treatments were described as immediate pain relief, close adhesion to the wound bed, diminished post-surgery discomfort, reduced infection rate, easiness of wound inspection (transparency), faster healing, and improved exudates retention, spontaneous detachment following re-epithelialization, and reduced treatment time and costs. There was one disadvantage of the limited elasticity in areas of great mobility. Gengiflex® was also developed to recover periodontal tissues. The achievements of both Biofill® and Gengiflex® in the application of bacterial cellulose (Cellumed) were applied in veterinary medicine to treat fresh, large surface wounds on horses and dogs [104, 105].

8.3.2.4 Bacterial cellulose as an artificial blood vessel

When the coronary vessels around the heart are not working well as a result of the hardening of the arteries, it might be necessary to carry out a bypass operation of vessels. Bacterial cellulose could be used for artificial blood vessels as it carries a lower risk of blood clots than the synthetic materials currently used for bypass operations [106, 107].

Bacterial cellulose can achieve a lower risk of blood clots than the synthetic materials currently in use. This means that the bacterial cellulose works very well in contact with the blood and is a very interesting alternative for artificial blood vessels. Real blood vessels have an internal coating of cells that ensure that the blood does not clot. This tube has an inner diameter of 1 mm, length of about 5 mm, and wall thickness of 0.7 mm. So these parameters are sufficient for experimental microsurgical requirements. A sufficient mechanical strength of the BASYC tubes is one of the essential properties for their use in microsurgery [108].

The material must resist both mechanical strains during microsurgical preparation and anatomizing and blood pressure of the living body. The native bacterial cellulose has mechanical properties, including shape retention and tear resistance, which are superior to many synthetic materials. In comparison with organic sheets, such as polypropylene, polyethylene-terephthalate or cellophane, bacterial cellulose processed into a film or sheet shows remarkable mechanical strength. Compliance mismatch between the synthetic graft and the surrounding native tissue has been reported as a major factor in ultimate failure of the currently used cardiovascular graft replacements. Thus, developing biomaterials that display close mechanical properties as the tissue it is replacing is an important objective in biomedical device design [109].

Polyvinyl alcohol (PVA) is a biocompatible hydrogel with characteristics desired for biomedical applications. It can be cross-linked by a low-temperature thermal cycling process. With the addition of a small amount of bacterial cellulose (BC) nanofibers, an anisotropic PVA-BC nanocomposite have been created by using a novel thermal processing method under an applied strain. The stress–strain tensile properties of porcine aorta were closely matched in both the circumferential and the axial directions by one type of anisotropic PVA-BC nanocomposite (10% PVA with 0.3% BC at 75% initial strain and cycle 2) within physiological range, with improved resistance to further stretch beyond physiological strains. The PVA-BC nanocomposite gives a broad range of mechanical properties, including anisotropy, by controlling material and processing parameters. PVA-BC nanocomposites with controlled degree of anisotropy that closely match the mechanical properties of the soft tissue it might substitute, ranging from cardiovascular to other connective tissues, can be obtained. On the other hand, bacterial cellulose has been carried out as a potential new vascular graft material by evaluating cell and blood interactions with bacterial cellulose.

The modification of bacterial cellulose with a novel technique, where xyloglucan is used as a carrier molecule for the adhesion-promoting peptide, resulted in increased cell adhesion, metabolism and cell spreading. Luminal coating of bacterial cellulose tubes with fibrin glue resulted in increased cell adhesion during static experiments and good cell retention during physiological shear stress.

"The evaluation of thrombogenicity in human blood plasma revealed that bacterial cellulose induces a slower coagulation compared with clinically available materials such as Gore-Tex® and Dacron®. The bacterial cellulose induced the least contact activation evaluated by XIIa generation. A Chandler loop system with freshly drawn blood reported that BC consumed low amounts of platelets and generated low thrombin values compared with Dacron® and Gore-Tex®" [110].

8.3.3 Applications of Bacterial Cellulose in Textile Sector

Bacterial cellulose has been found to combine some other organic or inorganic fibers to form various types of pellicle or film or non-woven fabrics. The microbial cellulose has been gaining interest in the production of medical dressing fabrics. It has been reported that bacterial cellulose fabric productions were investigated with its physical and mechanical properties of bacterial cellulose nonwovens and the possibility of producing inherently colored bacterial cellulose nonwovens by the addition of natural dyes. The weight, thickness, tensile load elongation percentage, tearing strength, and stiffness of bacterial cellulose showed that it was suitable for application as a fabric. It was also found that the tensile strength and elongation of bacterial cellulose fabrics were higher compared with most consumer nonwoven fabrics, while the tearing strength and stiffness of bacterial cellulose fabrics were similar to that of commonly used nonwoven fabrics. The weight of bacterial cellulose fabric was heavy due to the high water content in it [110]. The eco-friendly sustainable fabric showed the tensile strength that was two times stronger than top-grain leather; both have same appearance and thickness [111]. The integrated fabric was composed by culturing nine kinds of fabrics with bacterial cellulose producing bacteria. The bacterial cellulose was coated to the surfaces of sheets, and all become shinier [112].

8.3.4 Applications of Bacterial Cellulose in Environmental Treatment Sector

Effective recycling of the waste at source is one of the most efficient waste management methods. Bacterial cellulose can be used as a packaging material due to its biodegradability. Not only bacterial cellulose, but the substrate used for bacteria cellulose can also be reduced for tackling the environmental problem. Thus using this technology, coconut water generated at temples and damaged fruit waste can be recycled at source for minimizing environmental pollution with additional economic benefit and employment generation [3, 15].

It was reported that edible film can be an alternative to solve environmental problems. The products of edible film were created from different culture media, with different ratio of *Arenga pinnata* juice and coconut water. Their physical properties were tested and their biodegradation in soil was observed. The results showed that the percentage of weight loss increased during intervals of time. The highest percentage weight loss of edible film occurred with the media of 25% *Arenga pinnata* juice and 75% coconut water at the 9th day of cultivation [31].

Heavy metal, copper, and lead ion removal by carboxymethylated and phosphorylated bacterial cellulose has been reported [113, 114]. Therefore, bacterial cellulose is one of interesting materials for environmental treatment system.

8.3.5 Applications of Bacterial Cellulose in Paper Production Sector

Microbial cellulose has been reportedly used as a binder in papers, and because it consists of extremely small clusters of cellulose microfibrils, this property greatly adds to the strength and durability of pulp when integrated into paper [115]. Bacterial cellulose could be added into paper pulp to get better dry intensity, wet intensity, and higher water absorption at the same time solving the problem of removing lignin in traditional process of making paper. Not only the new paper but the recycled paper made from bacterial cellulose fibers also showed higher intensity [116].

Purified and dried bacterial cellulose was converted to a membrane to be used in the separation processes such as ultrafilteration, gas permeation, and vapor permeation and used in paper manufacture [70]. Ajinomoto Co., along with Mitsubishi Paper Mills in Japan, was currently active in developing microbial cellulose for paper products JP patent 63295793 [117]. Moreover, when the agitated bacterial cellulose and static bacterial cellulose were added at the wet end, both tensile strength and filler retention of the hand sheets were improved [118]. Especially, the bacterial cellulose from agitated culture had higher effects on filler-retention than that from static one. Thus, the bacterial cellulose may be useful as a wet-end additive for papermaking.

Furthermore, suspended solids in the nutrient medium for *Acetobacter xylinum* in a rotating-disk bioreactor become incorporated into the gelatinous mat of bacterial cellulose as it forms [117]. Embedding fibers of ordinary cellulose creates composites with enhanced strength and the toughness of bacterial cellulose. Purified cellulose and elongated fibers from paper are incorporated differently than are spherical particles such as silica gel.

About 90% of the final cellulose can come from scrap paper, and dried composite sheets were much stronger than plain bacterial cellulose per unit area. The incorporation of 5% of bacterial cellulose with wood pulp during paper sheet formation was found to significantly improve kaolin retention, strength, and fire resistance properties compared with the paper sheets produced from incorporating bacterial cellulose [119].

8.3.6 Applications of Bacterial Cellulose in Biocomposite Preparation

Bacterial cellulose can undergo cross-linking reaction, which is one mechanism by which many of the polymers, either naturally or synthetically produced, increase their potential range of applications; for example, cross-linking with fibrin in the presence of glutaraldehyde. The cross-link is formed between hydroxyl groups and amine groups found on bacterial cellulose and fibrin, respectively [120, 121].

Another study reported cross-linking between a composite comprising bacterial cellulose, alginate and calcium ions (Ca^{2+}) [122]. Bacterial cellulose composites are made by blending chitosan, poly (ethyleneglycol) (PEG), and gelatin for potential biomedical application of tissue-engineering scaffold and wound-dressing material [123].

Bacterial cellulose composites were successfully prepared by immersing a wet bacterial cellulose pellicle into chitosan, PEG, or gelatin solutions followed by freeze-drying. The products look like a foam structure. Scanning electron microscopy images show that chitosan molecules penetrated into bacterial cellulose forming a multilayer and a well-interconnected porous network structure with a large aspect surface. The morphology of the bacterial cellulose/gelatin scaffold indicates that gelatin molecules could penetrate well between the individual nanofibers of the bacterial cellulose. Cell adhesion studies for these composites were carried out using 3T3 fibroblast cells [124]. They showed much better biocompatibility than pure bacterial cellulose.

The inter-penetrated polymer network nanocomposite was obtained by the incorporation of poly(vinyl alcohol) (PVA) on the bacterial cellulose matrix by oven-drying. It is reported that both the addition of PVA and oven drying induce the appearance of larger pores (*circa* 1–3 μm in average diameter) in dried BC/PVA membranes. Both types of treatments also affect the permeability of the composite, as assessed by the diffusion coefficients of polyethylene glycol (PEG) molecules (900, 8,000, 35,000, and 100,000 Da) across the membranes. A mechanical property, Young's modulus of dry bacterial bacterial cellulose decreases following PVA incorporation, from 3.5 GPa to 1 GPa and a fivefold loss in tensile strength due to large porosity of composite membrane that are easier to rupture [125].

8.4 Conclusions

The production of bacterial cellulose is typically carried out by bacteria fermentation. The choice of strain, substrate, and medium design is particularly critical because it would affect the production and economic feasibility. Moreover, fermentation conditions such as fermentation method, pH, and temperature are critical for the optimal medium of fermentation.

The right choice of strain offers the advantage in terms of mechanical properties and water-holding capacity. The productivity of bacterial cellulose is a primary issue for its economic feasibility. Bacterial cellulose-based materials represent a new way of utilizing the inherent compatibility, environmental safety, and strength of blended natural cellulose. Under certain conditions, in particular after adding into foods, nata de coco may also give further appetency.

Moreover, bacterial cellulose has been used in the medical sector (i.e., wound-healing materials, drug delivery and other medical applications) environmental sector, textile, etc. In order for these materials to make transition into industrial technologies, a favorable cost/performance ratio is required.

Since bacterial cellulose is compatible with humans and is environment friendly, it can be applied to develop further innovative materials in near future.

Acknowledgements

The authors thank to Professor Hiroshi Tamura, Materials and Bioengineering Department, Kansai University, Japan, and Mr. Sularto S.Pd, Bionetech-lab, Indonesia, for their valuable suggestions on the whole of this study and Mr. Win Khaing Moe (Director General, Department of Research and Innovation, Ministry of Education, Myanmar) for his encouragement to fulfill of publication.

References

1. Brown, R. M. (2004). Cellulose structure and biosynthesis: What is in store for the 21th century, *J. Poly. Sci.*, **42**, pp. 487–495.
2. Panesar, P. S., Chavan, Y. V., Bera, M. B., Chand, O., and Kumar, H. (2009). Evaluation of acetobacter strain for the production of microbial cellulose, *Asian J. Chem.*, **21**(10), pp. S99–S102.
3. Rangaswamy, B. E., Vanitha, K. P., and Basavaraj-Hungund, S. (2015). Microbial cellulose production from bacteria isolated from rotten fruit, *Int. J. Polym. Sci.*, **2015**, pp. 1–8.
4. Park, J. K., Jung, J. Y., and Park, Y. H. (2003). Cellulose production by *Gluconacetobacter hansenii* in a medium containing ethanol, *Biotechnol. Lett.*, **25**, pp. 2055–2059.
5. Hestrin, S., and Schramm, M. (1954). Synthesis of cellulose by *Acetobacter xylinum*, *Biochem. J.*, **58**, pp. 345–352.
6. Chao, Y., Ishida, T., Sugano, Y., and Shoda, M. (2000). Bacterial cellulose production by *Acetobacter xylinum* in 50 L internal-loop air lift reactor, *Biotechnol. Bioeng.*, **68**, pp. 345–352.
7. Raghunathan, D. (2013). Production of microbial cellulose from the new bacterial strain isolated from temple wash waters, *Int. J. Curr. Microbiol. Appl. Sci.*, **2**(12), pp. 275–290.
8. Suwanposri, A., Yukphan, P., Yamada, Y., and Ochaikul, A. (2013). Identification and biocellulose production of *Gluconacetobacter* strains isolated from tropical fruits in Thailand, *Maejo Inter. J. Sci. Technol.*, **7**(01), 70–82.

9. Jagannath, A., Kalaiselvan, S. S., Manjunatha, P. S., Raju, and Bawa, A. S. (2008). The effect of pH, sucrose and ammonium sulphate concentrations on the production of bacterial cellulose (Nata-de-coco) by *Acetobacter xylinum, World J. Micro Biotechnol.*, **24**, pp. 2593–2599.
10. Sutanto (2012). Pineapple liquid waste as nata de pina raw material, *Makara Technol.,* **16**, pp. 63–67.
11. Ross, P., Mayer, R., and Benziman, M. (1991). Cellulose biosynthesis and function in bacteria, *Microbiol. Rev.*, **55**(1), pp. 35–58.
12. Zhao, H., Kwak, J. H., Zhang, Z. C., Brown, H. M., Arey, B. W., and Holladay, J. E. (2007). Studying cellulose fiber structure by SEM, XRD, NMR and acid hydrolysis, *Carbohydr. Polym.*, **68**, pp. 235–241.
13. Yamamoto, H., and Fumitaka, H. (1993). CP/MAS carbon-13 NMR analysis of crystal transformation induced for valonia cellulose by annealing at high temperatures, *Macromolecules,* **26**(6), pp. 1313–1317.
14. White, D. G., and Brown, R. M. (1989). Prospects for the commercialization of the biosynthesis of microbial cellulose, In: Schuerech, C. (ed), *Cellulose and Wood: Chemistry and Technology*, Wiley, New York.
15. Legge, R. L. (1990). Microbial cellulose as a specialty chemical, *Biotechnol. Adv.*, **8**, pp. 303–319.
16. Delmerl, D. P., and Amor, Y. (1995). Cellulose Biosynthesis, *Plant Cell.*, **7**, pp. 987–1000.
17. Sherif, M. A. S., and Sameshima, K. K. (2005). Evaluation of different carbon sources for bacterial cellulose production, *African J. Biotechnol.,* **4**(6), pp. 478–482.
18. Chawla, P. R., Ishwar, B. B., Shrikant, A. S., and Rekha S. S. (2009). Fermentative production of microbial cellulose, *Food. Technol. Biotechnol.*, **47**(2), pp. 107–124.
19. Muthukumarasamy, R., Revathi, G., Seshadri, S., and Lakshmi-narasimhan, C. (2002). *Gluconacetobacter diazotrophicus* (syn. *Acetobacter diazotrophicus*), a promising diazotrophic endophyte in tropics, *Curr. Sci.*, **83**(2), pp. 137–145.
20. Yu, X., and Atalla, R. H. (1996). Production of cellulose II by *Acetobacter xylinum* in the presence of 2,6-dichlorobenzonitrile, *Int. J. Biol. Macromol.*, **19**, pp. 145–146.
21. Iguchi, M., Yamanaka, S., and Budhiono, A. (2000). Bacterial cellulose–A masterpiece of nature's arts, *J. Mater. Sci.,* **35**, pp. 261–270.

22. Nadia H., Mohd-Cairulioqbal, H. A., and Ishak A. (2012). Physicochemical properties and characterization of *Nata de Coco* from local food industries as a source of cellulose, *Sains Malaysiana*, **41**(2), pp. 205–211.
23. Masaoka, S., Ohe T., and Sakota, N. (1993). Production of cellulose from glucose by *Acetobacter xylinum*, *J. Ferment. Bioeng.*, **75**, pp. 18–22.
24. Al-Shamary, E. E., and Al-Darwash, A. K. (2013). Influence of fermentation condition and alkali treatment on the porosity and thickness of bacterial cellulose membranes, *Online J. Sci. Technol.*, **3**(2), pp. 194–203.
25. Changjin, S., Seonyong, C., Jieun, L., and Seongjun, K. (2002). Isolation and cultivation characteristics of *Acetobacter xylinum* KJ-1 producing bacterial cellulose in shaking cultures, *J. Microbiol. Biotechnol.*, **12**(5), pp. 722–728.
26. Andelib, A. Y., and Nuran, D. A. (2009). Isolation of cellulose producing bacteria from wastes of vinegar fermentation, *Proceedings of the World Congress on Engineering and Computer Science 2009,* WCECS, San Francisco, USA, **1**, pp. 6–8.
27. Kamarudin, S., Sahaid, K., Sobri, M., Mohtar, W., Radiah, D., and Norhasliza, H. (2013). Different media formulation on biocellulose production by *Acetobacter xylinum* (0416), *J. Sci. Technol.*, **21**(1), pp. 29–36.
28. Afreen, S. S., and Lokeshappa, B. (2014). Production of bacterial cellulose from *Acetobacter xylinum* using fruits wastes as substrate, *Int. J. Sci. Technol.,* **2**(8), pp. 57–64.
29. Santos, S. M., Carbajo, J. M., and Villar, J. C. (2013). The effect of carbon and nitrogen sources on Bacterial cellulose production and properties from *Gluconacetobacter Sucrofermentans* CECT7291 focused on its use in degraded paper restoration, *Bio. Reso.*, **8**(3), pp. 3630–3645.
30. Keshk, S. (2002). *Gluconacetobacter xylinus*: A new resource for cellulose, *Egypt J. Biotechnol.*, **1**, pp. 305–311.
31. Faridah, Yusrini, M., Zahra, F., and Rahmayani (2010). The effect of culture medium on bacterial cellulose production as edible film material, *Chem. Eng. Dep.*, pp. 20–24.
32. Hungund., Prabhu, S., Shetty, C., Acharya, S., Prabhu, V., and Gupta, S. (2013). Production of Bacterial cellulose from *Gluconacetobacter persimmonis* GH-2 using dual and cheaper carbon sources, *J. Micro. Biochem. Technol.*, **5** pp. 031–033.

33. Matsuoka, M., Tsuchida, T., Mastushita, K., Adachi, F., and Yoshinaga (1996). A synthetic medium for bacterial cellulose production by *Acetobacter xylinum Subsp. Sucrofermentans, Biosci. Biotech. Biochem.*, **60**(4), pp. 575–579.

34. Hestrin, S., and Schramm, M. (1954). Synthesis of cellulose by *Acetobacter xylinum*: preparation of freeze-dried cells capable of polymerizing glucose to cellulose, *Biochem J.*, **58**, pp. 345–352.

35. Esin, P. C., and Halil, B. (2011). Evaluation of different pH and temperatures for bacterial cellulose production in HS (Hestrin-Scharmm) medium and beet molasses medium, *Afri. J. Microbiol. Res.*, **5**(9), pp. 1037–1045.

36. Kouda, T., Yano, H., and Yoshinaga, F. (1997). Effect of agitator configuration bacterial cellulose productivity in aerated and agitated culture, *J. Ferment. Bioeng.*, **83**, pp. 371–376.

37. Takayasu, T., and Fumihiro, Y. (1997). Production of bacterial cellulose by agitation culture systems, *Pure. Appl. Chem.*, **69**(11), pp. 2453–2458.

38. Son, H. J., Kim, H. G., Kim, K. K., Kim, H. S., and Kim, Y. G. (2003). Increased production of bacterial cellulose by *Acetobacter* sp. V6 in synthetic media under shaking culture conditions, *Bioresour. Technol.*, **86**, pp. 215–219.

39. Pourramezan, G. Z., Roayaei, A. M., and Qezelbash, Q. R. (2009). Optimization of culture conditions for bacterial cellulose production by *Acetobacter* sp. 4B-2, *Biotechnology*, **8**, pp. 150–154.

40. Hwang, J. W., Yang, Y. K., Hwang, J. K., Pyun Y. R., and Kim, Y. S. (1995). Effects of pH and dissolved oxygen on cellulose production by *Acetobacter xylinum* BRC5 in agitated culture, *J. Biosci. Bioeng.*, **88**(2), pp. 183–188.

41. Galas, E., Krystynowicz, A., Tarabasz-Szymanska, L., Pankiewictz, R., and Zyska, M. (1999). Optimization of the production of bacterial cellulose using multivariable linear regression analysis, *Acta. Biotechnol.*, **19**, pp. 251–260.

42. Sasithorn, K. (2008). Bacterial cellulose production by *Acetobacter xylinum* strains from agricultural waste products, *Appl. Biochem. Biotechnol.*, **148**(1–3), pp. 245–256.

43. Phunsri, A., Tammarate, P., Krusong, W., and Tantratian, S. (2003). The liquid/air interface area and depth of liquid medium suitable for cellulose production from *Acetobacter* TISTR 975, *J. Sci. Res. Chula. Uni.*, **28**, pp. 35–43.

44. Gromet-Elhanan, Z., and Hestrin, S. (1963). Synthesis of cellulose by *Acetobacter xylinum*. VI. Growth on citric acid-cycle intermediates, *J. Bacteriol.*, **85**, pp. 284–292.

45. Lacadin, A. L., and Millan, E. R. (1980). Relative effectiveness of coconut, dalanghita, incale and tomato as media for producing nata, *TCA. Res. J.*, **2**(3), pp. 235–245.

46. Shoda, M., and Sugano, Y. (2005). Recent advances in bacterial cellulose production, *Biotechnol. Bioprocess. Eng.*, **10**, pp. 1–8.

47. Brown, Jr. R. M. (1987). The biosynthesis of cellulose, *Food Hydrocolloids*, **1**, pp. 345–351.

48. Okiyama, A., Shirae, H., and Kano, H. (1992). Bacterial celluolse I, two-stage fermentation process for cellulose production by *Acetobacter aceti*, *Food Hydrocolloids,* **6**(5), pp. 471–477.

49. Pacheco, J., Yee, S., Zentella, M., and Marvan, E. (2004). Cellulosa bacterianaen *Gluconacetobacter xylinum*: Biosintesisy aplicaciones, *Revista Especializada en Ciencias. Quimico – Biolôgicas,* **7**(1), pp. 18–25.

50. Geyer, U., Heinze, T., and Stei, A. (1994). Formation, derivatization and applications of bacterialcellulose, *Int. J. Biol. Macromol.,* **16**(6), pp. 343–347.

51. Keshk, S. M. A. S., and Sameshima, K. (2006). Utilization of sugar cane molasses with/without the presence of lignosulfonate for the production of bacterial cellulose, *Appl. Microb. Biotech.*, **72**, pp. 291–296.

52. Denise, M., Rosilene, A., Adenise, L., and Gilvan, W. (2011). Application of bacterial cellulose conservation of minimally processed fruits, *Revista Brasileira. De. Tecnologia. Agroindustrial.,* **5**(1), pp. 356–366.

53. Nasab, M. M., and Yousefi, A. R. (2010). Investigation of physicochemical properties of the bacterial cellulose produced by *Gluconacetobacter xylinus* from date syrup, *World. Acad. Sci. Eng. Technol.*, **44**, pp. 08–26.

54. Bielecki, S., Krystynowicz, A., Turkiewicz, M., and Kalinowska, H. (2005). Bacterial cellulose, in *Polysaccharides and Polyamides in the Food Industry*, Steinbüchel, A., and Rhee, S. K. (eds.), Wiley-VCH Verlag, Weinheim, Germany, pp. 31–85.

55. George, J., Ramana, K. V., Sabapathy, S. N., Jambur, H. J., and Bawa, A. S. (2005). Characterization of chemically treated bacterial (*Acetobacter xylinum*) biopolymer: Some thermo-mechanical properties, *Int. J. Biol. Macromol.,* **37**, pp. 189–194.

56. Ogawa, R., and Tokura, S. (1992). Preparation of bacterial cellulose containing N-acetylglucosamine residues, *Carbohydr. Polym.*, **19**, pp. 171–178. Yoshino, T., Asakura, T., and Toda, K. (1996). Cellulose production by *Acetobacter pasteurianus* on silicone membrane, *J. Ferment. Bioeng.*, **81**, pp. 32–36.

57. Jonas, R., and Farah, L. H. (1998). Production and application of microbial cellulose, *Poly. Deg. Stab.*, **59**, pp. 101–106.

58. Shirai, A., Takahashi, M., Kaneko, H., Nishimura, S., and Ogawa, M. (1994). Biosynthesis of novel polysaccharide by *Acetobacter xylinum*, *Int. J. Biol. Macromol.*, **16**, pp. 297–300.

59. Cannon, R. E., and Anderson, S. M. (1991). Biogenesis of bacterial cellulose, *Crit. Rev. Microbiol.*, **17**, pp. 435–447.

60. Hwang, J. W., Yang, Y. K., Hwang, J. K., Pyun, Y. R., and Kim, Y. S. (1999). Effects of pH and dissolved oxygen on cellulose production by *Acetobacter xylinum* BRC5 in agitated culture, *J. Biosci. Bioeng.*, **88**, pp. 183–188.

61. David, N. S. (1996). *Chemical modification of Lignocellulosic Materials: Chemical Structures of Cellulose, Hemicelluloses and Lignin*, Marcel Dekker. Inc., New York, USA.

62. Novaes, A. B., and Novaes, A. B. Jr. (1993). Bone formation over a TiAl6V4 (IMZ) implant placed into an extraction socket in association with membrane therapy (Gengiflex), *Clin. Oral. Implant. Res.*, **4**, pp. 106–110.

63. Bäckdahl, H., Helenius, G., Bodin, A., Nannmark, U., Johansson, B. R., and Risberg, B. (2006), Mechanical properties of bacterial cellulose and interactions with smoth muscle cells, *Biomaterial,* **27**(9), pp. 2141–2149.

64. Wanichapichart, P., Kaewnopparat, S., Buaking, K., and Puthai, W. (2002). Characterization of cellulose membranes produced by *Acetobacter xylinum*, *J. Sci. Technol.*, **24**, pp. 855–862.

65. Kuan, C., Jeffrey, M., and Ali, D. (2009). Effect of different additives on bacterial cellulose production by *Acetobacter xylinum* and analysis of material property, *Cellulose,* **16**, pp. 1033–1045.

66. Luan, J., Wu, J., Zheng, Y., Song, W., Wang, G., Guo, J., and Ding, X. (2012). Impregnation of silver sulfadiazine into bacterial cellulose for antimicrobial and biocompatible wound dressing, *Biomed. Mater.*, **7**, doi:10.1088/1748-6041/7/6/065006.

67. Yamanaka, S., Watanabe, K., and Suzuki, Y. (1990). *European patent* 0,396,344,A2.

68. Andersson, J., Stenhamre, H., Bäckdahl, H., and Gatenholm, P. (2010). Behavior of human chondrocytes in engineered porous bacterial cellulose scaffolds, *J. Biomed. Mater. Res. Part A.*, **94**, pp. 1124–1132.

69. Muhammad Mustafa, A., Mohd-Cairul, I., Mohd-Amina., and Claire, M. (2014). Review of BC-based drug delivery systems, Royal Pharmaceutical Society, *J. Pharma. Pharmacol.*, **66**, pp. 1047–1061.

70. Dos-Anjos, B., Novaes, A. B., Meffert, R. Jr., and Barboza, E. P. (1998). Clinical comparison of cellulose and expanded polytetrafluoroethylene membranes in the treatment of class II furcations in mandibular molars with 6-month re-entry, *J. Periodontol.*, **69**, pp. 454–459.

71. Krystynowicz, A., and Bieleck, S. (2001). Biosynthesis of bacterial cellulose and its potential application in the different industries, *Polish. Biotetechnol. News.*, Copyright by Bio–Tech Consulting.

72. Klemm, D., Schumann, D., Udhardt, U., and Marsch, S. (2001). Bacterial synthesized cellulose–artificial blood vessels for microsurgery, *Polym. Sci.*, **26**(9), pp. 1561–1603.

73. Luz, D., Carreno, P., Luis, A., and Alberto, C. (2006). Effect of culture and purification conditions on physicochemical and transport properties in bacterial cellulose membranes, *Biomaterials*, ***27***, pp. 145–151.

74. Fontana, J. D., Souza, A. M., Fontana, C. K., Toriani, I. L., and Moreschi, J. C. (1990). Acetobacter cellulose pellicle as a temporary skin substitute, *Appl. Biochem. Biotechnol.*, **24–25**, pp. 253–264.

75. Czaja, W. K., Young, D. J., Kawecki, M., and Brown, R. M. Jr. (2007). The future prospects of microbial cellulose in biomedical applications, *Biomacromolecules*, **8**, pp. 1–12.

76. Andrade, F. K., Silva, J. P., Carvalho, M., Coutinho, E. M. S., Soares, R., and Gama, F. M. (2011). Studies on the hemocompatibility of bacterial cellulose, *J. Biomed. Mater. Res. Part A*, **98**, pp. 554–566.

77. Backdahl, H., Risberg, B., and Gatenholm, P. (2011). Observations on bacterial cellulose tube formation for application as vascular graft, *Mater. Sci. Eng. C*, **31**, pp. 14–21.

78. Alvarez, O. M., Patel, M., Booker, J., and Markowitz, L. (2004). Effectiveness of a biocellulose wound dressing for the treatment of chronic venous leg ulcers: Results of a single center randomized study involving 24 patients, *Wounds*, **16**, pp. 224–233.

79. Klemm, D., Heublein, B., Fink, H., and Bohn, A. (2005). Cellulose: Fascinating biopolymer and sustainable raw material, *Angew. Chem. Int. Ed. Eng.*, **44**, pp. 3358–3393.

80. Czaja, W., Krystynowicz, A., Bielecki, S., and Brown, R. M. (2006). Microbial cellulose-the natural power to heal wounds, *Biomaterials*, **27**, pp. 145–151.

81. Putra, A., Kakugo, A., Furukawa, H., Gong, J. P., and Osada, Y. (2008). Tubular bacterial cellulose gel with oriented fibrils on the curved surface, *Polymer*, **49**, pp. 1885–1991.

82. Trovatti, E., Freire, C. S., Pinto, P, C., Almeida, I. F., Costa, P., Silvestre, A. J., Neto, C. P., and Rosado, C. (2012). Bacterial cellulose membranes applied in topical and transdermal delivery of lidocaine hydrochloride and ibuprofen: In vitro diffusion studies, *Int. J. Pharm.*, **435**, pp. 83–87.

83. Lin, Z., Guan, Z., and Huang, Z. (2013). New bacterial cellulose/ polyaniline nanocomposite film with one conductive side through constrained interfacial polymerization, *Ind. Eng. Chem. Res.*, **52**, pp. 2869–2874.

84. Stoica-Guzun, A., Stroescu, M., Tache, F., Zaharescu, T., Grosu, E. (2007). Effect of electron beam irradiation on bacterial cellulose membranes used as transdermal drug delivery systems, *Nucl. Instrum. Meth. Phys. Res.,* **265**, pp. 434–438.

85. Yoon, S. H., Jin, A. J., Kook, M. C., and Pyun, Y. R. (2006). Electrically conductive bacterial cellulose by incorporation of carbon nanotubes, *Biomacromolecules,* **7**, pp. 1280–1284.

86. Jung, R., Kim, Y., Kim, H. S., and Jin, H. J. (2009). Antimicrobial properties of hydrated cellulose membranes with silver nanoparticles, *J. Biomater. Sci. Polym. Ed.,* **20**, pp. 311–324.

87. Pinto, R. J., Marques, P. A., Neto, C. P., Trindade, T., Daina, S., and Sadocco, P. (2009). Antibacterial activity of nanocomposites of silver and bacterial or vegetable cellulosic fibers, *Acta Biomater.,* **5**, pp. 2279–2289.

88. Maneerung, T., Tokura, S., and Rujiravanit, R. (2007). Impregnation of silver nanoparticles into bacterial cellulose for antimicrobial wound dressing, *Carbohydr. Polym.*, doi:10.1016/j. carbpol. 2007.07.025.

89. Habibi, Y., Lucia, L. A., and Rojas, O. J. (2010). Cellulose nanocrystals: Chemistry, self-assembly, and applications, *Chem. Rev.*, **110**, pp. 3479–3500.

90. Wei, B., Yang, G., and Hong, F. (2011). Preparation and evaluation of a kind of bacterial cellulose dry films with antibacterial properties, *Carbohydr. Polym.,* **84**, pp. 533–538.

91. Bodhibukkana, C., Srichana, T., Kaewnopparat, S., Tangthong, N., Bouking, P., Martin, G. P., and Suedee, R. (2006). Composite membrane of bacterially-derived cellulose and molecularly imprinted polymer for use as a transdermal enantioselective controlled-release system of racemic propranolol, *J. Control. Release.*, **113**, pp. 43–56.

92. Suedee, R., Bodhibukkana, C., Tangthong, N., Amnuaikit, C., Kaewnopparat S., and Srichana, T. (2008). Development of a reservoir-type transdermal enantioselective-controlled delivery system for racemic propranolol using a molecularly imprinted polymer composite membrane, *J. Control. Release.*, **129**, pp. 170–178.

93. Fang, J., Sung, K., Wang, J. J., and Chen, C. C. (2002). The effects of iontophoresis and electroporation on transdermal delivery of buprenorphine from solutions and hydrogels, *J. Pharm. Pharmacol.*, **54**, pp. 1329–1337.

94. Fang, J. Y., Hwang, T. L., and Leu, Y. L. (2003). Effect of enhancers and retarders on percutaneous absorption of flurbiprofen from hydrogels, *Int. J. Pharm.*, **250**, pp. 313–325.

95. Mori, R., Nakai, T., Enomoto, K., Uchio, Y., and Yoshino, K. (2011). Increased antibiotic release from a bone cement containing bacterial cellulose, *Clin. Orthopol. Relat. Res.*, **469**, pp. 600–606.

96. Biondi, M., Ungaro, F., Quaglia, F., Netti, P. A. (2008). Controlled drug delivery in tissue engineering, *Adv. Drug. Deliv. Rev.*, **60**, pp. 229–242.

97. Ougiya, H., Watanabe, K., Morinaga, Y., and Yoshinaga, F. (1997). Emulsion effect of bacterial cellulose, *Biosci. Biotech. Biochem.*, **61**, pp. 1541–1545.

98. Nitar, N., Tetsuya, F., and Hiroshi, T. (2010). Selection of a biopolymer based on attachment, morphology and proliferation of fibroblast NIH/3T3 cells for the development of a biodegradable tissue regeneration template: Alginate, bacterial cellulose and gelatin, *Proc. Biochem.*, **45**, pp. 457–466.

99. Li, J., Wan, Y., Li, L., Liang, H., and Wang, J. (2009). Preparation and characterization of 2,3-dialdehyde bacterial cellulose for potential biodegradable tissue engineering scaffolds, *Mater. Sci. Eng.*, **29**, pp. 1635–1642.

100. Chiaoprakobkij, N., Sanchavanakit, N., Subbalekha, K., Pavasant, P., and Phisalaphong, M. (2011). Characterization and biocompatibility of bacterial cellulose/alginate composite sponges with human

keratinocytes and gingival fibroblasts, *Carbohydr. Polym.*, **85**, pp. 548–553.

101. Svensson, A., Nicklasson, E., Harrah, T., Panilaitis, B., Kaplan, D. L., Brittberg, M., and Gatenholm, P. (2005). Bacterial cellulose as a potential scaffold for tissue engineering of cartilage, *Biomaterials*, **26**, pp. 419–431.

102. Kawecki, M., Krystynowicz, A., Wysota, K., Czaja, W., and Sakiel, S. (2004). Bacterial cellulose biosynthesis, properties and applications, *Int. Re. Conf. Biotechnol.*, Vienna, Austria, pp. 14–18.

103. Shibazaki, H., Kuga, S., and Fumihiko, O. (2003). Bacterial cellulose membrane as seperationmedium, *J. Appl. Poly. Sci.*, **50**(6), pp. 965–969.

104. Fontana, J. D., Franco, V. C., Souza, S. J., Lyra, I. N., and Souza, A. M. (1991). Nature of plant stimulators in the production of *Acetobacter xylinum* (tea fungus) biofilm used in skin therapy, *Appl. Biochem. Biotechnol.*, **28–29**, pp. 341–351.

105. Farah, L. F. X. (1990). Process for thepreparation of cellulose film, cellulose film produced thereby artificial skin graft and its use, *BIO FILL PRODUTOS BIOTECNOLÓGICOS (BRASIL).*, US 4912049.

106. Bodin, A., Backdahl, H., Fink, H., Gustafsson, L., Risberg, B., and Gatenholm, P. (2007). Influence of cultivation conditions on mechanical and morphological properties of bacterial cellulosetubes, *Biotechnol. Bioeng.*, **97**, pp. 425–434. DOI: 10.1002/bit.21314.

107. Klemm, D., Udhardt, U., Marsch, S., and Schumann, D. (1999). BASYC–Bacterial synthesized cellulose: Miniaturized tubes for microsurgery, *Cellulose Polymer News*, **24**(11), pp. 373–380.

108. Keshk, S. M. A. S. (2014). Bacterial cellulose production and its industrial applications, *J. Bioproc. Biotech.*, **4**(2), pp. 1–10, doi: 10.4172/2155-9821.1000.

109. Kowalska-Ludwicka, K., Cala, J., Grobelski, B., Sygut, D., Jesionek-Kupnicka, D., Kolodziejczyk, M., Bielecki, S., and Pasieka, Z. (2013). Modified bacterial cellulose tubes for regeneration of damaged peripheral nerves, *Arch. Med. Sci.*, **9**, pp. 527–534. DOI: 10.5114/aoms.2013.33433.

110. Wood, D., Liu. H., and Carol, J. S. (2015). Production and characterization of bacterial cellulose fabrics, *Annu. Conf. Proc. Inter. Tex. Apparel Asso.*, ITAA, Paper 143.

111. Mizuno, M., Kamiya, Y., Katsuta, T., and Amano, Y. (2012). Creation of bacterial cellulose- fabric complexed material, *Fiber*, **68**, pp. 42–47.

112. Yim, S. M. Song, J. E., and Kim, H. R. (2016). Production and characterization of bacterial cellulose fabrics by nitrogen sources of tea and carbon sources of sugar, *Process. Biotecnol.*, htpp://dx.doi.org/10.1016/j.procbio.2016.07.001.

113. Chen, S., Zou, Y., Yan, Z., Shen, W., Shi, S., Zhang, X., and Wang, H. (2009). Carboxymethylated bacterial cellulose for copper and lead ion removal, *J. Hazard Mater.*, **161**, pp. 1355–1359.

114. Oshima, T., Kondo, K., Ohtoaba, K., Inoue, K., and Baba, Y. (2008). Preparation of phosphorylated bacterial cellulose as an adsorbent for metal ions, *React. Funct. Polym.*, **68**, pp. 376–383.

115. Surma, S. B., Presler, S., and Danielewicz, D. (2008). Characteristics of bacterial cellulose obtained from *Acetobacter xylinum* culture for application in papermaking, *Fibres. Textil. Eastern. Europe*, **16**(**4**(69)), pp. 108–111.

116. Neelobon, S., Jiraporn, B., and Suwannee, T. (2007). Effect of culture conditions on bacterial cellulose (BC) production from *Acetobacter xylinum* TISTR 976 and physical properties of BC parchment paper, *Suranaree J. Sci.-Technol.*, **14**(4), pp. 357–365.

117. Hioki, N., Hori, Y., Watanabe, K., Morinaga, Y., and Yoshinaga, F. (1995) Bacterial cellulose; as a new material for papermaking, *Kami-Gikyo-Shi.*, **49**, pp. 82–87.

118. Basta, A. H., and El-Saied, H. (2009). Performance of improved bacterial cellulose application in the production of functional paper, *J. Appl. Microbiol.*, **107**, pp. 2098–2107.

119. Lauro, X. F., Pierre B., and Carlos, V. C. (2007). Biopaper obtained from microorganisms, *Proc. the 1st Inter. Workshop on Adv. in Cleaner Production*, November 21–23, Sao Paulo, SP, Brazil.

120. Klemm, D., Schmauder, H. P., and Heinze, T. (2002). *Biopolymers*, Weinheim, Wiley, VCH.

121. Shi, Z., Phillips, G. O., and Yang, G. (2013). Nanoscale electroconductive composites, *Nanoscale*, **5**, pp. 3194–3201.

122. Charreau, H., Foresti, M. L., and Vazquez, A. (2013). Nanocellulose patents trends: A comprehensive review on patents on cellulose nanocrystals, microfibrillated and bacterial cellulose, *Recent Pat Nanotechnol.*, **7**, pp. 56–80.

123. Li, M., Tian, X., and Chen, X. (2009). Modeling of flow rate, pore size, and porosity for the dispensing-based tissue scaffolds fabrication, *Manuf. Sci. Eng.*, **131**(3), pp. 34501–34505.

124. Brown, E. E., Laborie, M.-P. G., and Zhang, J. (2012). Glutaraldehyde treatment of bacterial cellulose/fibrin composites: Impact on morphology, tensile and viscoelastic properties, *Cellulose,* **19**, pp. 127–137.

125. Alexandre, F. L., João, P. S., Fernando, D., and Miguel, G. (2013). Production and characterization of a new bacterial cellulose/ poly (vinyl alcohol) nanocomposite, *Materials*, **6**, pp. 1956–1966; doi:10.3390/ma6051956.

Chapter 9

Thermal, Mechanical and Degradation Properties of Starch-Based Bio-Nanocomposites

Mohan Turup Pandurangan and Krishnan Kanny

Composite Research Group (CRG), Department of Mechanical Engineering, Durban University of Technology, Durban, South Africa

mohanp@dut.ac.za

9.1 Introduction

Starch is a white tasteless carbohydrate that is sourced from natural materials, which include fruits, vegetables, seeds, roots, tubers and stem pith (the soft spongy tissue in the centre of most vascular plants) of plants. It consists of two types of molecules: the linear and helical amylose and the branched amylopectin. It is notably found in potatoes, corn, wheat and rice. However, as humans we cannot consume all of the starch due to the cellulose found in starch [1]. We do not have the enzymes to break down the cellulose. The basic chemical formula of a starch molecule is $C_6H_{10}O_6$ as shown in Fig. 9.1.

Handbook of Biopolymers: Advances and Multifaceted Applications
Edited by Shakeel Ahmed, Suvardhan Kanchi, and Gopalakrishnan Kumar

ISBN 978-981-4800-17-4 (Hardcover), 978-0-429-02475-7 (eBook)
www.panstanford.com

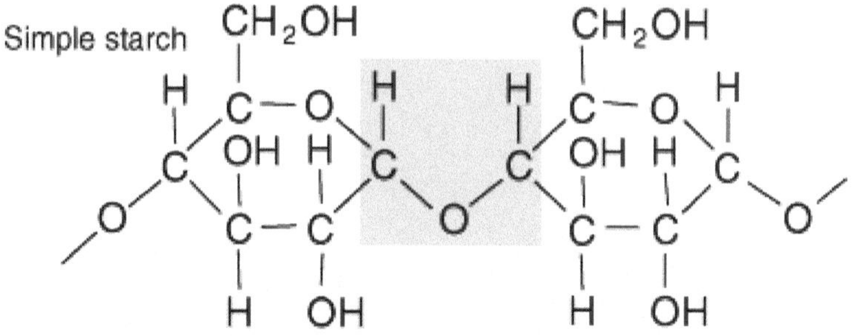

Figure 9.1 Chemical structure of starch [2].

Starch-based polymers are environmentally friendly and biodegradable but possess several challenges for applications such as low thermal and mechanical properties, hydrophilic characteristics with low barrier properties and quicker biodegradation. In this chapter, we will examine the effect of modifying the starch-based biopolymer using nanoparticle addition as a filler material. We will use a series of nanoparticles fillers such as carbon nanotubes, nanoclay and nano-oxide/carbide materials on the starch-based polymer to examine its effect on thermal, mechanical, barrier and degradation properties on the polymer [3].

9.2 Starch-Based Bio-Nanocomposites

9.2.1 Starch Bio-Nanocomposites Filled by Whiskers

Whiskers are typically cellulose nanocrystals. They are a linear condensation polymer consisting of β-1,4-linked D-anhydroglucopyranose units (AGU) (Fig. 9.2). Whiskers can be isolated from their original biomass like flax, hemp, ramie and wood, through acid hydrolysis with concentrated mineral acids under strictly controlled conditions of time and temperature [4]. Acid action results in a decrease of the amorphous parts by removing polysaccharide material closely bonded to the crystallite surface and breaking down portions of glucose chains in most accessible, nanocrystalline regions and by acid hydrolysis of cellulosic materials. The resulted rods usually have lengths ranging from 100–500 nm and diameters ranging from 10 to

30 nm [5]. The hydrolysis time can affect the structure and properties of the resulted nanocrystals (length, L; diameter, D; and L/D value) and thus on the structure and performance of the nanocomposites [6].

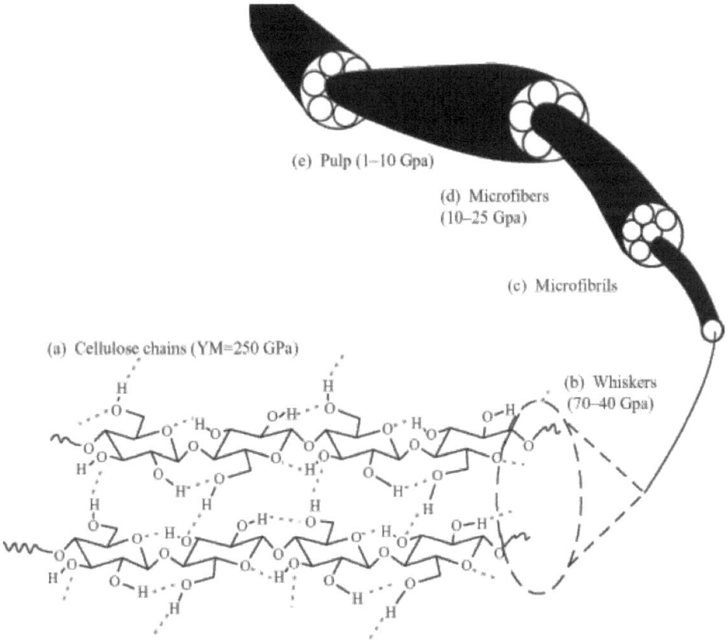

Figure 9.2 Schematic representation of cellulose chemical structure and microfibril formation [7].

In other research, tunicin (sea animal cellulose) whiskers were also utilized, which are slender, parallel-piped rods of 500 nm to 1–2 μm length and 10 nm width, into plasticized matrixes [8, 9, 36]. The high aspect ratio (length to diameter) of these nanoparticles (50–200) increases the interfacial phenomena between whiskers and matrix. A series of studies were conducted of plasticized waxy maize starch/tunicin whiskers nanocomposites. The results indicated that the strong interactions between whiskers such as hydrogen bonds, which could be affected by moisture content, led to a rigid network, which is responsible for the improved mechanical performance. However, it was reported that the trans-crystallization phenomenon of amylopectin chains on the palatino mechanical properties of the ensuing nanocomposites.

9.2.2 Starch Bio-Nanocomposites Filled by Starch Nanocrystals

Waxy maize starch nanocrystals were obtained by acid hydrolysis of native granules by strictly controlling the temperature of the process, the acid and starch concentrations, the hydrolysis duration and the stirring speed [10]. Waxy maize starch nanocrystals consist of 6–8 nm-thick, platelet-like particles with a length of 20–40 nm and a width in the range 15–30 nm. Figure 9.3 shows the TEM micrographs of negatively stained waxy maize starch nanocrystals for different hydrolysis times. One of the advantages of these nanocrystals is that their chemical structures are similar to those of starch matrix and thus there is an excellent compatibility between filler and matrix. In Fig. 9.3a, a lamellar organization is clearly revealed with the platelets lying parallel to the incident electron beam. In Figs. 9.3b,c, parallelepipedal platelets are seen lying flat on the carbon film. The arrow in Fig. 9.3b indicates a pyramidal stack of crystals. Figure 9.3d shows nearly individual waxy maize starch nanocrystals obtained after 6 weeks of hydrolysis (scale bars: 50 nm).

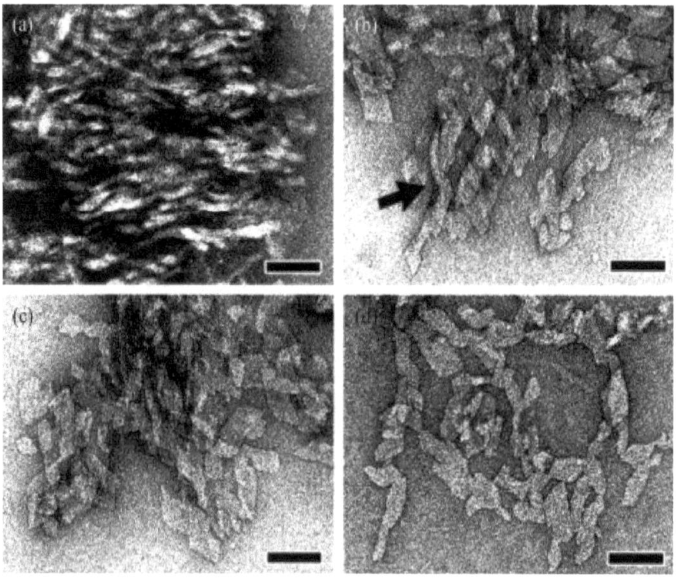

Figure 9.3 TEM micrographs of waxy maize starch samples [14].

The waxy maize starch nanocrystals were used as a reinforcing agent in a waxy maize starch matrix plasticized with glycerol or sorbitol [11]. Results have shown that the reinforcing effect of starch nanocrystals can be attributed to strong filler/filler and filler/matrix interactions due to the establishment of increased hydrogen bonding. The presence of starch nanocrystals led to the reduction in the molecular mobility of amylopectin chains at room temperature and thus to a considerable slowing down of the recrystallization of the matrix during aging in high humidity atmospheres [12]. This could result in a slower evolution of the mechanical properties of the filled materials compared to the unfilled matrix. The nanocomposites can also show a decrease in water vapour permeability, which is attributed to the well dispersion of the nanocrystals introducing a tortuous route for water molecules to pass through the films [13].

9.3 Structure and Morphology of Starch-Based Bio-Nanocomposites

Cellulose nanofibrils are 3–10 nm-thick fibrils formed during cellulose biosynthesis in higher plants, whereas the microfibrillated cellulose nanofiber consists of aggregates of elementary fibrils. Figures 9.4a–d show different morphologies of starched-based composites [15, 20]. The structure of the cellulose fibrils is stringy and does not have a broad width compared to the rest. This could result in the fibril having low tensile properties. The cellulose crystals, on the other hand, have short strands that are more consistent and would possess higher compression properties. If you compare the morphology of the cellulose fibrils and bacterial cellulose, you can notice that they are very similar in structure. Now looking at the morphology of starch nanocrystals, you can notice that they are completely different from the rest; the width is much thicker than the others and they are not string-like structures [21].

Nanofibrillated or microfibrillated cellulose is isolated from the cell wall of lignocellulosic materials using a high mechanical sheer force. The cellulose nanofibers are gathered by the disintegration of the plant fibre cell wall and they generally have nanometric length and width of micrometres and possess a high specific surface area. The cellulose nanofibre is a highly attractive organic biodegradable reinforcement in polymer composites, due to its good inherent mechanical properties, high speed ratio and its ability to form a network [22]. For fibrillation of plant fibres different technologies have been used to employ mechanical treatments such as high-pressure homogenizing [23], cryo-crushing [24], and ultrasonication [25]. Very often, pre-treatment methods such as enzymatic methods [26] and chemo-mechanical methods [27] are used to facilitate the fibrillation process.

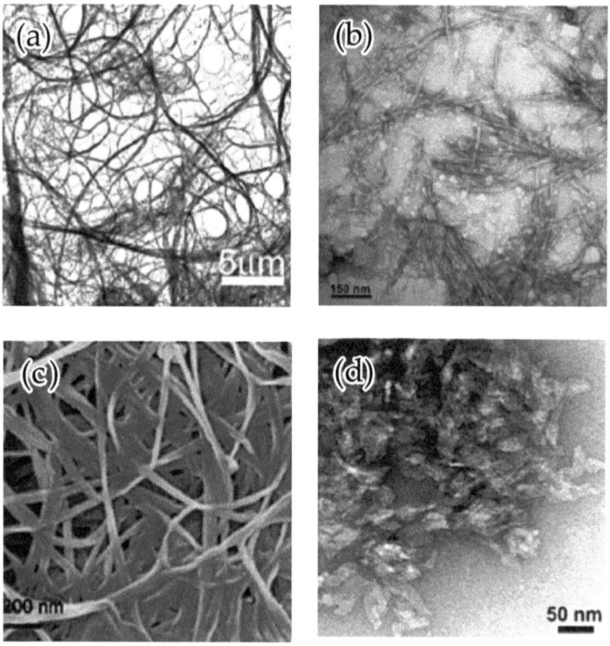

Figure 9.4 (a) morphology of cellulose fibrils [16]; (b) morphology of cellulose nanocrystals [17]; (c) morphology of bacterial cellulose [18]; (d) morphology of starch nanocrystals [19].

9.4 Mechanical and Thermal Properties of Starch-Based Materials

Starch is one of the most studied natural and renewable polymers with many promising applications in biodegradable packaging. During the past decade, starch-based nanocomposites with the addition of low quantities of inorganic or organic fillers have given rise to significant improvements in physical, mechanical, and barrier properties. This section will focus on the mechanical and thermal properties of starch-based nanocomposites considering starch as matrix. It will discuss some examples of starch nanocomposites reinforced with polysaccharide-based (cellulose, starch) fillers, and attempt to give a critical comparison of the different reinforcements.

9.4.1 Botanical Origin: Amylose/Amylopectin Ratio

The botanical origin, especially the amylose/amylopectin ratio as shown in Table 9.1 is known to influence the mechanical properties of starch. The idea generally spread is that amylose-rich materials display higher strength and lower strain at break than amylopectin rich materials. However, this affirmation should be nuanced due to some opposite results described in the literature. Results showed that above 20 wt.% glycerol the strain at break was more important for amylose-rich films [28]. In fact, two behaviours can be reported according to the plasticizer content [29].

Table 9.1 Amylose and amylopectin content of starch from different sources [32]

Type of starch	Amylose (%)	Amylopectin (%)
Amylomaize	48–77	23–52
Banana	17–24	76–83
Corn	17–25	75–83
Cush-cush Yam	9–15	85–91
Chickpeas	30–40	60–70
High-amylose com	55–70	30–45
Potato	17–24	76–83
Rice	15–35	65–85
Sorghum	25	75
Tapioca (cassava)	19–22	28–81
Wheat	20–25	75–80
Waxy	<1	>99

Figure 9.5 shows the molecular structure of amylose and amylopectin. For low plasticizer contents, an increase in the amylose/amylopectin ratio would improve both the strength and the strain at break because of the linear structure of amylose chains. For high plasticizer contents, an increase in the amylose content results in a mechanical reinforcement (the strength increases and the strain decreases). This last tendency is related to fast amylose crystallization in comparison with slower amylopectin retrogradation. In any case, it can be concluded that the amylose/amylopectin ratio is not a determining parameter for the improvement of ultimate mechanical properties [31].

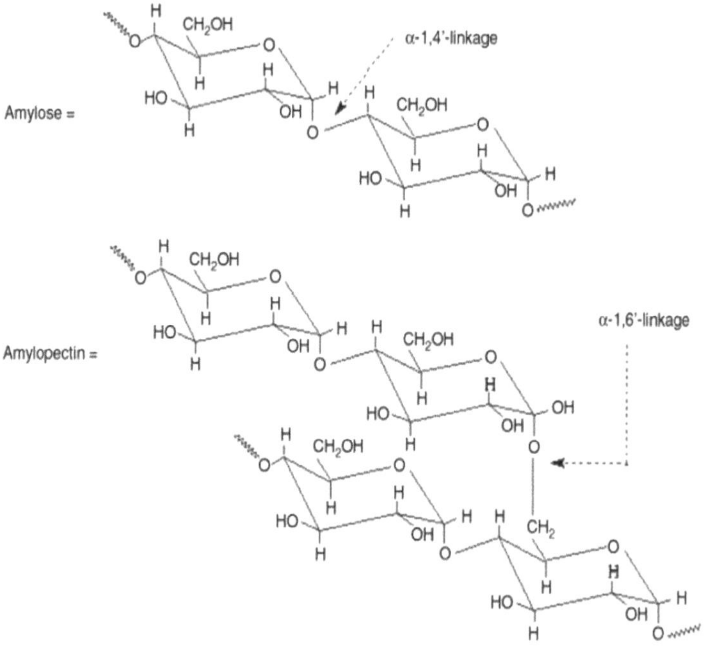

Figure 9.5 Molecular structures of amylose and amylopectin [30].

9.4.2 Plasticization

Starch is most often plasticized with water and polyols such as glycerol. For example, a glycerol content of 25% allows decreasing the glass transition temperature of potato starch-based materials stored at 57% relative humidity (RH) from 285°C to −10°C. Depending on the plasticizer content, two mechanical behaviours are evidenced irrespective of the starch origin. For a low plasticizer content, mechanical properties present similarities to the anti-plasticization effect generally observed in synthetic polymers; that is, both the stress and strain at break decrease when the plasticizer concentration increases to up to 12 wt.% [33]. This phenomenon would correspond to a local macromolecular motion limitation due to strong interactions between starch chains and plasticizer resulting in a reinforcing hydrogen network. For higher amounts of plasticizer, mechanical properties are controlled by the glass transition, which involves the molecular motions. The classical plasticization effect is characterized by an increase in the strain at break and a decrease in both the strength and Young's modulus. Plasticizers (water and glycerol) play the role of internal lubricants, preventing the rigidification of non-crystallized macromolecular starch chains at ambient temperature [34]. The plasticization effect is the phenomenon that governs the mechanical properties of unaged systems. Currently, all formulations of thermoplastic starch materials require the use of very large amounts of plasticizer (roughly 20 wt.%) to avoid ant plasticization events. As a result, starch-based materials have globally poor mechanical properties due to the relaxation of the polymer network in the presence of plasticizer. In the case of plasticized starch/glycerol systems, dynamic mechanical analysis (DMA) highlights the presence of two relaxation phenomena associated with a two-step storage modulus drop. It is characterized by two peaks on loss modulus and tan δ curves, around −40 and 40°C (25 wt.% of glycerol and 15 wt.% of water), revealing the heterogeneity of the system and a phase separation. The first peak is associated with the relaxation of glycerol-rich domains, and the second one with starch-rich domains [35].

9.4.3 Aging

Physical aging, i.e. retrogradation, induces structural changes without chemical modification. This phenomenon is slow because it generally occurs at temperatures below the glass transition temperature of starch-based materials (T_g) for which the chain mobility is reduced. It only occurs at room temperature for plasticized systems displaying a T_g below room temperature. Starch-based materials in the rubbery state are the site of recrystallization, which takes place in two steps: a fast recrystallization step during film processing where gelatinized starch recrystallizes according to the B-type, and a slow recrystallization upon film storage [36]. Mechanical properties of starch-based materials are directly affected by physical aging [37]. It inevitably leads to higher strength and Young's modulus and lower strain at break irrespective of the starch origin [38]. Crystallites play the role of physical cross-linking bonds, resulting in stress concentration and thus making more brittle the material [39]. Therefore, the mechanical properties of starch-based materials are very sensitive to the ambient conditions and moisture content, which are difficult to control.

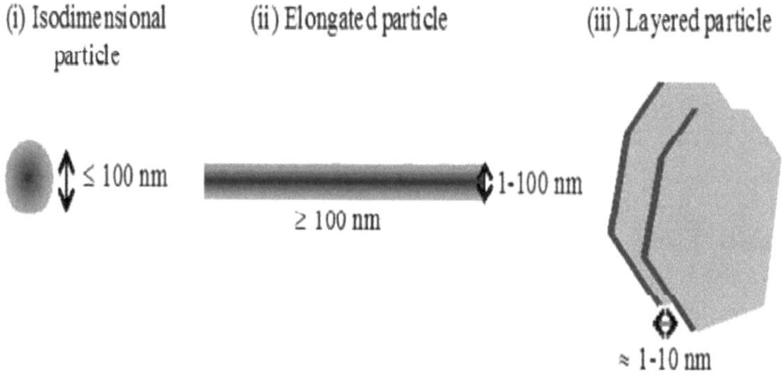

Figure 9.6 Schematic representation of the three categories of fillers [40].

9.4.4 Fillers

Fillers can be classified into three major categories according to their particle size and shape as shown in Fig. 9.6. (i) Iso-dimensional particles have the same size in all directions and their aspect ratio is usually close to unity (examples include spherical silica, carbon black, and fullerenes). This type of particles has not yet been used as fillers in a starch matrix, probably due to their moderate reinforcing effect. (ii) Elongated particles consist of fibrils with diameters ranging between 1 and 100 nm and lengths of several hundred of nanometres. They included cellulose and chitin nanofibrils and carbon nanotubes [41]. These nanocomposites are known to present better mechanical properties due to the higher particle aspect ratio. (iii) The third category is composed of layered particles such as layered silicates, but also starch nanocrystals. They are characterized by one dimension, often the thickness, ranging from several angstroms to several nanometres. The most common fillers used as reinforcing agents in a starch matrix are layered silicates and carbon nanotubes, which are both inorganic fillers, and polysaccharide-based (cellulose, chitin, starch) nanocrystals, which are organic fillers [42].

Polysaccharides (cellulose, chitin, starch) are among the most promising sources to produce nanoparticles with various morphologies and geometrical characteristics providing a wide range of nanoparticle properties. As layered silicates, polysaccharides offer significant potential for surface modification using well-established carbohydrate chemistry, allowing researchers to tailor the surface functionality of the nanoparticles [43].

9.4.4.1 Cellulose nanocrystals

Aqueous suspensions of cellulose nanocrystals can be prepared by acid hydrolysis of native substrates. The object of this treatment is to dissolve away regions of low lateral order so that the water-insoluble, highly crystalline residue may be converted into a stable suspension. The resulting nanocrystals occur as rod-like particles or whiskers, whose dimensions depend on the nature of the substrate and preparation conditions. They are typically a few hundred nanometres long and in the order of few nanometres in diameter as shown in Table 9.2 [44]. The aspect ratio of these whiskers is defined as the ratio of the length to the diameter.

Table 9.2 Geometrical characteristics of rod-like cellulose nanofibers: length (L) and diameter (D) [45]

Source of cellulose	L (mm)	D/W (nm)
Potato pulp	>1000	2–4
Tunicate	500–2000	10
Cottonseed linter	170–490	40–60
Cotton	100–300	5–10
Wheat straw	100–300	30–40
Hemp/pea hull	240–400	7–12
Cassava bagasse	360–1700	2–11

9.4.4.2 Starch nanocrystals

Starch can also be used as a source to produce nanocrystals. In Fig. 9.7, it can be seen how the starch nanocrystals are obtained from mild acid hydrolysis of native starch granules using hydrochloric or sulphuric acid. The use of sulphuric acid leads to more stable aqueous suspensions due to resulting negatively charged surfaces.

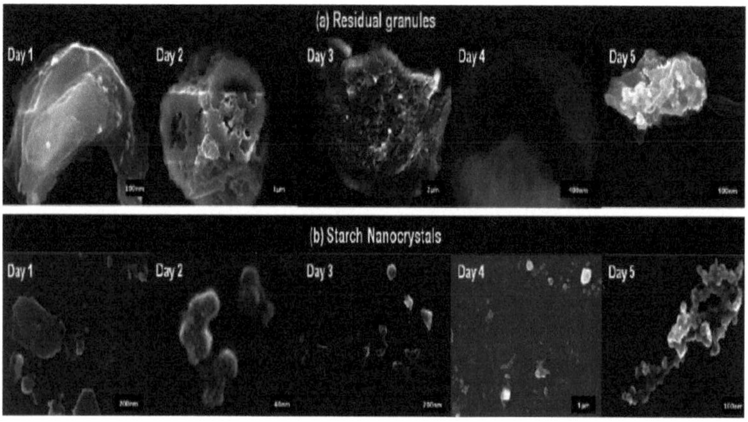

Figure 9.7 SEM of starch during the acid hydrolysis process [46].

Starch nanocrystals correspond to the highly crystalline water-insoluble lamellae of the native starch granules. Three main botanical origins were used for that purpose: waxy maize, pea, and potato [47]. Starch nanocrystals generally appear as aggregates of parallelepiped platelets exhibiting 60–65° acute angles. The botanical origin of starch, namely the crystallinity type and the relative proportion of amylase and amylopectin, determines the structure and the morphology of starch nanocrystals. The conditions of acid hydrolysis also affect the yield and the final dimensions of starch nanocrystals. The dimensions of starch nanocrystals have been characterized by image analysis of transmission electron microscopy (TEM) pictures, with a length ranging from 20 to 150 nm, width from 10 to 30 nm, and thickness from 5 to 10 nm, depending on the botanical origin [48]. It has been recently reported that at any time, both microscale and nanoscale particles can be found and coexist in suspensions [49]. Cross-flow filtration was proved to be an efficient continuous operation for separating the nanoparticles from the bulk suspension [50]. Starch nanocrystals, as cellulose whiskers, possess a highly reactive surface covered with hydroxyl groups, which provides the possibility of modification via a chemical reaction strategy. The purpose of the chemical modification is to contribute to specific functions and to expand the applications of starch nanocrystals. There are three strategies for chemical modification of starch nanocrystals: modification by chemical reaction with small molecules, grafting of polymer chains using coupling agents and grafting of polymer chains by polymerization of a monomer [51].

9.4.5 Mechanical Properties of Nanocomposites Based on Starch and Organic Reinforcements

The reinforcing effect of cellulose microfibrils, is obtained from potato tuber cells, in a glycerol plasticized potato starch-based matrix. Above T_g, a great increase in the storage modulus (more than 2 decades) and in the thermal stability was observed even with a filler content of as low as 5 wt.%. With increasing filler content, no marked evolution of the rubbery modulus was noted, but the terminal zone occurred at higher temperatures [52]. Mechanical properties of highly plasticized materials were strongly dependent on relative humidity conditions. The high dependence of mechanical properties of starch/cellulose-based nanocomposite materials on plasticizer and moisture content was greater than glycerol and sorbitol plasticized waxy maize starch reinforced with tunicin whiskers.

In the case of glycerol-plasticized systems, the reinforcing effect of tunicin whiskers is very low as compared to experimental data obtained for other matrices such as poly (S-*co*-BuA) [53]. The reinforcing effect of cellulose whiskers is generally ascribed to the presence of strong interactions between whiskers such as hydrogen bonds, which lead to the formation of a rigid network governed by the percolation threshold, besides the geometry and stiffness of the whiskers. Thus, it was deduced that plasticizers (glycerol and water) probably hindered the formation of such hydrogen-bonded cellulose network within the starch-based matrix due to an accumulation of plasticizer in the cellulose/ amylopectin interfacial zone. This phenomenon was also reported for glycerol/sorbitol-plasticized cassava starch/cassava bagasse cellulose nanofibers systems in Table 9.3 [54]. However, the reinforcing effect of tunicin whiskers in sorbitol-plasticized waxy maize starch was higher than in corresponding glycerol-plasticized systems due to the absence of trans crystallization of sorbitol at the cellulose/amylopectin interface, as proved by DSC and WAXS studies. Thus, sorbitol seems to be more adequate to be used as plasticizer [55].

Table 9.3 Relative rubbery storage tensile modulus (E_R) and relative tensile test characteristics (nominal Young's modulus, E nominal stress at break, σ_R: nominal strain at break, ε_R) for starch-cellulose whiskers nanocomposites [56]

	Cellulose	Filler content					
Starch matrix	**origin**	**RH(%)**	**(Wt.%)**	E'_R	E_R	σ_R	ε_R
Potato starch/ 30 wt% glycerol	Potato pulp	0	5	125	—	—	—
			10	150	—	—	—
Potato starch/ 30 wt% glycerol	Potato pulp	58	7.5	—	18	—	—
		15	—	25	—	—	—
			23	—	40	—	—
			31	—	50	—	—
			39	—	60	—	—
Waxy maize starch/33 wt% glycerol	Tunicate	58	3.2	0.3	15	8	1.6
			6.2	16	15	8	1.3
			16.7	1	2.5	8	1.2
Waxy maize starch 33 wt% sorbitol	Tunicate	58	5	1	1.2	1.8	1
			10	2.5	3.6	4	1
			15	2.5	5.3	4.3	1
			25	3.5	7.2	10.3	1
Wheat starch 42 wt% glycerol	Cottonseed linter	50	2.5	2.2	1.6	1.2	0.9
			5	3	2.5	1.7	0.85
			10	5.2	4.3	2.1	0.8
			20	10	8.3	2.6	0.65
			30	15	10	3.1	0.45

Starch matrix	Cellulose origin	Filler content RH(%)	(Wt.%)	E'_R	E_R	σ_R	ε_R
Wheat starch	Cotton (from What man filter paper)	50	2.1	—	3.6	—	1.4
			5	—	6.7	—	3.1
			10.3	—	9	—	3.3
Modified potato starch	Wheat straw	50	2	1.55	1.4	1.2	—
			5	1.8	1.75	1.35	—
			10	2.75	2.45	1.75	—
Pea starch/ 36 wt% glycerol	Hemp	43	5	—	1.1	1.15	0.85
			10	—	3.5	1.55	0.75
			15	—	5.3	1.75	0.45
			20	—	7.6	1.9	0.3
			30	—	25.8	2.95	0.1
Pea starch/ 30 wt% glycerol	Pea hull	43	10	—	—	1.95	0.85
Cassava starch/30 wt% glycerol	Cassava bagasse	53	5	1.1	1.4	1.4	2.6
			10	2.1	1.5	1.6	2.6
			20	2.1	1.6	1.4	1.8
Maize starch/ 30 wt% glycerol	Wheat straw	75	5	1.05	1.4	1.1	—
			10	2.1	2.3	1.2	—
			15	3.4	2.95	1.4	—

Cellulose whiskers are stiff and straight well-defined objects, whereas cellulose microfibrils are flexible and hairy fibres. Therefore, in composites filled with cellulose microfibrils, the reinforcing effect occurs most probably from both the formation of the hydrogen-bonded cellulose network and a tangling effect. In composites filled with cellulose whiskers, only the former phenomenon is likely occurring.

The low reinforcing effect of cellulose whiskers in glycerol-plasticized systems was not unanimously reported. For example, in the case of glycerol-plasticized wheat starch/cottonseed linter whiskers, potato starch or maize starch/wheat straw nanofibers, or pea starch/hemp nanofibers nanocomposite materials, Young's modulus, the tensile strength, and the storage modulus at the rubbery state were significantly improved with increasing filler content shown in Table 9.4. These results indicated an efficient stress transfer from matrix to filler [57], which was attributed to strong interfacial interactions between the cellulose crystallite surface and the starch matrix, and to the formation of a rigid network of nanofibers connected by hydrogen bonds. The shift of T_g to higher temperatures (about 30–40°C) higher in the case of glycerol plasticized potato starch/wheat straw nanofibers systems confirmed this hypothesis. For some systems, the elongation at break decreased at the same time [58]. However, the reinforcing effect of cellulose whiskers is not necessarily accompanied by a decrease in the elongation at break, as observed for different glycerol plasticized systems.

Table 9.4 Tensile properties of corn starch-based nanocomposite films with different plasticizers at the same level (15 wt.%) [59]

	Tensile strength (MPa)	**Elongation at break (%)**
TPS	14.22 ± 0.98	5.26 ± 0.83
Glycerol-plasticized TPS/MMT	15.60 ± 0.63b	4.44 ± 0.52c
Urea-plasticized TPS/MMT	21.19 ± 2.69b	2.49 ± 0.55d
Formamide-plasticized TPS/MMT	26.64 ± 3.02a	3.25 ± 0.59d

Note: Mean±standard deviation of each analysis; means with the same letters are not significantly different (*P* # 0.05).

9.4.6 Thermal Properties

From thermomechanical measurements based on dynamical mechanical thermal analysis (DMTA) characterization, we can determine the influence of the nanofillers on the local mobility of the chains and thus on their relaxation temperatures.

Past studies have shown that there were two main relaxation peaks for both plasticized starch and TPS/MMT nanocomposites. The high-temperature transition was the α-transition (glass transitions, T_g) of starch matrix; and the low-temperature transitions was generally regarded as the β-transition (secondary transition, T_β) of the plasticizer like glycerol. These two relaxation peaks were shifted towards higher temperatures on increasing MMT clay content, indicating that the starch molecular chain in nanocomposites was restricted by the MMT exfoliated nanostructure, and thus the degree of freedom decreased [60]. Other studies found that MMT increased the two relaxation peaks while other organoclays like Cloisite 30B, 10A and 6A decreased them. In addition, the nanostructure of TPS/MMT composites could also be evidenced by the increased storage modulus from DMTA measurement. These results were attributed to the matching of the surface polarity and interactions of MMT and TPS [61]. However, in a study of the effect of type of organoclays on the glass transition temperature of starch acetate nanocomposite foams, we found that all organoclays (Cloisite 30B, 10A, 25A, and 20A) increased the glass transition temperature of the nanocomposites, which was attributed to the restriction of the movement of starch acetate segments imposed by the nanodispersion of starch acetate molecules in silicate layers [62].

The thermal stability of starch/clay nanocomposites has been studied by thermogravimetric analysis (TGA) in detail. Although little or no improvement of thermal stability of starch/nanoclay composites has been reported in a few studies, it has been more widely suggested that the addition of nanoclay can enhance the thermal stability of the nanocomposites [64]. By investigating the thermal stability of TPS/MMT nanocomposites with different clay content in Fig. 9.8, we can conclude that clay can act as a heat barrier, which enhances the overall thermal stability of the nanocomposites. The inorganic material has better thermal stability and resistance than the organic material due to its chemistry; thus, the incorporation of inorganic particles improved the thermal stability of starch [65]. Other results also showed that the TPS/MMT and TPS/Cloisite 30B hybrids have a higher degradation temperature than the neat matrix. The largest increase for both kinds of nanocomposites was observed with clay content of 5 wt.%, whereas the increase diminished with further increasing clay content. Also, the thermal stability of TPS/MMT nanocomposites was higher than that of the TPS/Cloisite 30B, regardless of the clay content. Such a result highlighted a relationship between the clay dispersion state and the thermal stability [66]. Besides, MMT possesses a high aspect ratio (width to thickness) and can act as a barrier to the volatile products during heating [67]. There may be three factors determining the behaviour of clay-filled starch at thermal treatment, which can be written in the following order of effectiveness of samples during thermal treatment: clay dispersion, hydroxyl group exposure and re-association of matrix chains. Thus, higher clay dispersion, lower concentration of exposed hydroxyl groups and fewer chances of starch chain re-association are favourable for increasing the thermal decomposition temperature [68].

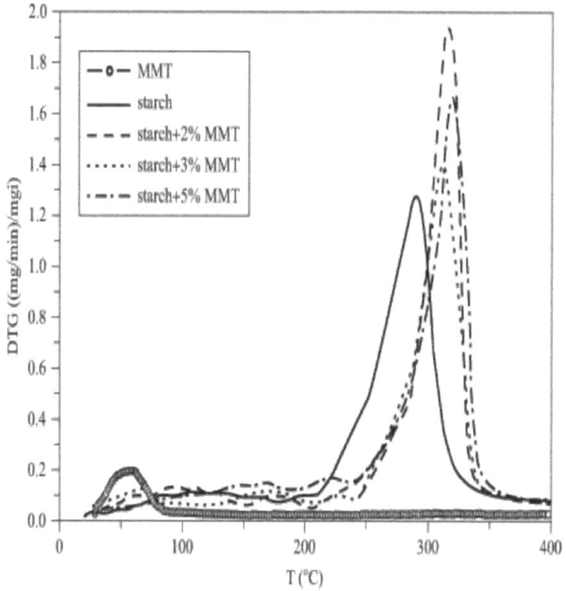

Figure 9.8 Thermogravimetric curves of films of starch, MMT and its nanocomposites [63].

9.5 Barrier and Biodegradation Properties

9.5.1 Barrier Properties

A packaging material should have the necessary barrier properties against water, water vapour, gases/air and grease/oils depending on the end use application, with the aim to protect the material from ambient environment or prevent the loss of flavours, fragrance, and moisture from food products. Any single material layer used in the packaging is able to provide a barrier against water, or water vapour, or gas, or grease, or combinations of maximum two or three properties, but it rarely provides full protection against all four properties. Therefore, combinations with other materials in multi-layer packaging or coatings are applied [69].

It has been widely reported that the addition of nanoclay to TPS can reduce the water vapour absorption/permeability (WVA/WVP) of the resulted hybrids. The moisture absorption (rate) of TPS/MMT nanocomposites was effectively reduced compared with that of TPS as shown in Fig. 9.9. Better moisture barrier properties were obtained with increased clay content.

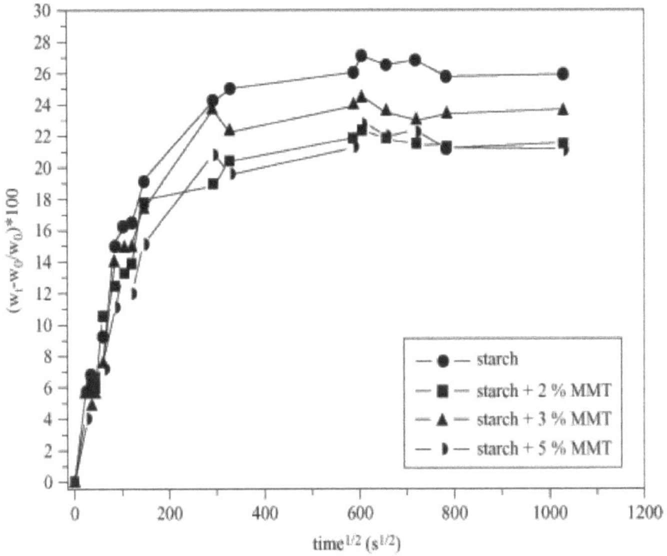

Figure 9.9 Water vapour and oxygen permeability [70].

Biopolymers are used as packaging material, but are mostly blended with other biopolymers to form biocomposites with improved barrier properties [71]. However, no biopolymer or biocomposite coating was found to share three or four barrier properties at the same time. It seems that the utilization of biopolymers or biocomposites reached a saturated state. Nanotechnology could open up the gates for further developments towards the utilization of biopolymers as barrier coatings. For example, biopolymers can be either applied as nanoparticles over paper to improve water resistance or incorporated with nanoscale additives such as nanocellulose, nanoclays, inorganic and organic nanoparticles to enhance the barrier properties. As such, nanotechnology can bring additional water resistance to biopolymer and bio-nanocomposite coatings, which would benefit further to maintain the existing water vapour and gas permeability of composites and not deteriorate them in presence of high relative humidity or with direct water contact. The nanocellulose components in form of micro- and nanofibrillated cellulose (NFC/MFC) or cellulose nanocrystals (CNC) also allow to provide specific gas and oil barrier properties in addition to good mechanical properties when used in native form (e.g. carboxymethylated NFC or c-NFC) added as a bio-based filler in the biopolymer matrix (e.g. PLA-NFC). Nanotechnology also enables the better valorization of resources in the pulp and paper industry, as the nanocellulose can be produced from recovered material fractions from generated wastewaters [72].

9.5.2 Biodegradation Properties

Degradable biopolymers are polymers which undergo a significant change in chemical structure under specific environmental conditions. These changes result in a loss of physical and mechanical properties. Biodegradable biopolymers undergo degradation from the action of naturally occurring microorganisms such as bacteria, fungi, and algae. These polymers may also be designated as photodegradable, oxidatively degradable, hydrolytically degradable, or those which may be composted. Biopolymers have the capability to be degraded through the action of naturally occurring organisms leaving behind organic by-products as CO_2 and H_2O with no toxic or no environmentally harmful residue [73].

Natural filler materials may be incorporated into synthetic plastic matrices as a rapidly biodegradable component. Often, granular starch is added to polyethylene to increase the degradation rate of the plastic material. Starch can also be used in its gelatinized form [74]. Heating the starch in the presence of water during extrusion or injection moulding causes the formation of a thermoplastic material that may be deformed during blending. This starch-based product is then blended with either natural or synthetic materials. Heating starch above its glass transition temperature breaks its molecular structure, allowing further bonding [75]. Glycerol is often used as a plasticizer in starch blends, to increase softness and pliability. Starch granules that have been plasticized with water and glycerol are referred to as plasticized starches [76]. Polymer materials that are formed from starch-based blends may be injection moulded, extruded, blown, or compression moulded. Agricultural feedstocks for the biopolymer industry also include fibres that are used as reinforcing fillers. This classification includes cellulose, which is the highly polar, main structural component of flax and hemp fibres [77]. Natural cellulose fibres are low cost and biodegradable and have strong mechanical properties. These characteristics make cellulose fibres the most common choice for natural fillers in plastic materials [78]. The presence of 25% of cellulose fibres in a polypropylene matrix causes a significant increase in tensile modulus. Cellulose has a very long molecular chain, which is infusible and insoluble in all but the most aggressive solvents [79]. Therefore, it is most often converted into derivatives to increase solubility, which further increases adhesion within the matrix. Flax fibres continue to receive most of the consideration, as they are mechanically strong and readily available. Chemical treatment (acetylation) of the fibres is performed in order to modify the surface properties, without changing the fibre structure and morphology [80]. These modifications slow down the initiation of degradation of the fibres, and increase adhesion at the fibre and matrix interface.

From Fig. 9.10, we established from weight loss over time that starch biodegrades at a faster rate than its clay composites.

Fibres that have been thoroughly dried prior to being added to the matrix show improved adhesion as opposed to fibres with a higher moisture content. Research has shown that polyvinyl alcohol is an appropriate polymer to use as a matrix in natural fibre reinforced composites, as it is highly polar and biodegradable [82].

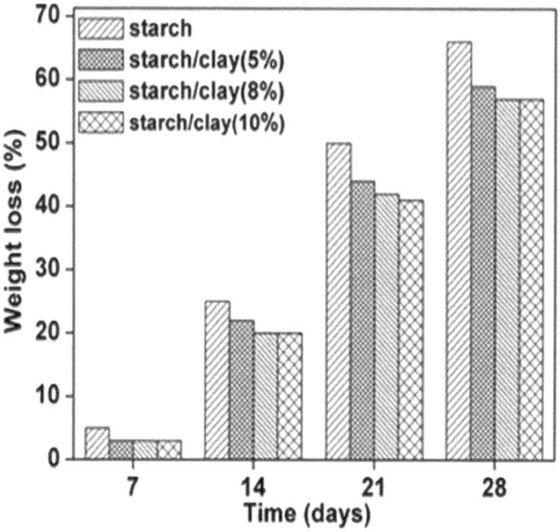

Figure 9.10 Biodegradability of starch and the starch/clay bio-nanocomposites of 5%, 8% and 10% clay concentrations [81].

9.6 Processing and Product Development of Starch-Based Bio-Nanocomposites

9.6.1 Applications of Starch-Based Bio-Nanocomposites

Processing nanocomposites results much more complex than neat starch due to the inherent complications related to the nanofillers use. Its high surface/volume ratio makes the electrostatic interactions between them, such as Van der Waals interactions, maximized, generating many clusters which are too difficult to break during processing. Furthermore, most of the employed loads in starch composites are hydrophilic and therefore show strong hydrogen bridge interactions, contributing to the agglutination problem. For that reason, different strategies have been developed to incorporate nanofillers to the matrix, with a special focus on a good dispersion achievement [83].

In the next sections, we briefly discuss the processing of nanocomposites by the casting and extrusion techniques.

9.6.1.1 Casting

Casting is one of the most common techniques for processing lab-scale starch composites. Different types of polysaccharide nanofillers such as cellulose from flax, wood, hemp ramie, cassava bagasse, wheat straws, starch from waxy maize, regular maize, chitin and chitosan, among others, can be used to fabricate starch nanocomposites by the solution casting method.

The main drawback associated with this method is that polysaccharide nanofillers tend to cling together due to the strong hydrogen bonding interactions between them (especially cellulose nanowhiskers (C-NW)) [84].

In general, for the dispersion of these fillers in starch solutions, some additional treatments, such as ultrasonication, sonication and homogenization, are required.

In some cases, special considerations need to be considered. Starch nanoparticles (S-NP), for example, must be added at reduced temperature to avoid the gelatinization of their amorphous component and the destruction of their special structure, which affects their reinforcing ability [85]. A possible attempt to solve this problem is to modify starch nanoparticles with citric acid by dry preparation technique. These modified nanoparticles dispersed in distilled water were added before gelatinization at 90°C. In this case it is reported that S-NP could not be swelled or gelatinized in hot water due the cross-linking induced by the citric acid [86].

In the case of starch or cellulose nanocrystals obtained by the sulphuric acid treatment, and prepared in aqueous condition, the suspension experiences neither sedimentation nor flocculation, because of charge repulsion due to the sulphate groups created on their surface. This feature makes it easier to add them into to the starch solution.

The way in which different nanofillers are added into the matrix varies according to different authors. A nanoparticle suspension of waxy maize starch can be added to gelatinized starch, but they do not specify the suspension preparation protocol. Adding the suspension of waxy maize starch nanocrystals in the desired quantities to gelatinized matrix of cassava starch, and then the mixture must be stirred for 10 min at 250 rpm and degassed for another 1 h. A rice starch solution must be prepared with a concentration of 3% (w/v) in distilled water. Then heat the dispersion with continuous stirring until it was completely gelatinized (85°C for 5 min), and finally cool it to 45 ± 2°C [87].

Subsequently, previously lyophilized nanocrystals were dispersed in water and incorporated to the gelatinized starch suspension. By contrast, some authors add the suspension together with the starch before gelatinization. A TPS nanocomposite reinforced with cellulose nanofibers (C-NF) extracted from whear straw was prepared by this method. They shear mixed maize starch and glycerol (30%) for 10 min using a Fluko FA25 homogenizer (10,000 rpm) and prepared a C-NF water dispersion using a bath sonicator for 3 h. The C-NF dispersion was added to the starch–glycerol mixture and the final product was subjected to further shear mixing for 20 min [88]. Finally, the starch + glycerol + C-NF mixture was continuously stirred (at 80–100 rpm) using a mechanical stirrer and heated at 75±3°C.

Citric acid-modified starch nanoparticles (CA-S-NP) can be added into a water, glycerol, and starch mixture, prior to gelatinization [89].

The glycerol content was fixed at 36 wt.% based on the dry starch matrix. Then this mixture was poured into a flask equipped with a stirrer and heated at 100°C for 30 min to gelatinize starch.

To summarize, in the preparation of starch-based nano-composites by casting technique, it is generally necessary to consider how to prevent the nanoparticle clumping [90].

9.6.1.2 Extrusion

The problems associated with starch extrusion are increased when starch nanocomposites are processed using hydrophilic fillers. As it was explained before, although these fillers are compatible with starch and glycerol, they tend to clump together obstructing their dispersion in the matrix [91].

Some authors propose the preparation of a dry starch, glycerol, and nanofiller mixture without water addition (only the starch moisture is used) to be extruded, while others underline the problem of the hydrophilic fillers and propose their incorporation into water suspensions, either by pumping the solution to the molten polymer into a subsequent extrusion stage or by feeding the extruder with the mixture of all the components with water [92]. Two types of food extruders are used for processing of prepared blends of starchy components mixed with glycerol added up to 20% in weight; functional additives like talc, polyvinyl alcohol, and foaming agent (PLASTRONFOAM PDE). The additives were added up to 3% in weight, while water up to 5% in weight. The blends were extruded at temperature range 80–170°C, variable screw rotational speed up to 150 rpm, and die pressure of 12 to 18 MPa [93]. A premixed starch, plasticizer, lubricant, and cellulose nanofiber (C-NF) gels with high water contents and then extruded the obtained dispersions. The authors prepared nanocomposite with 0, 5, 10, 15, and 20 wt.% of cellulose nanofiber. To remove the excess water, an extruder barrel equipped with two atmospheric vents and vacuum ventilation can be used [94].

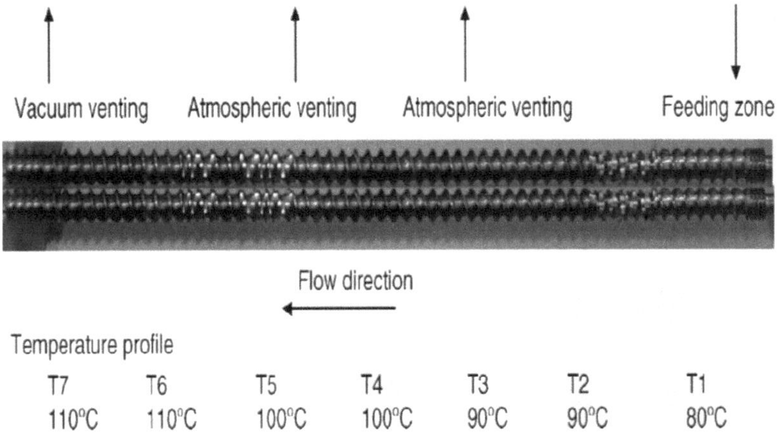

Figure 9.11 Screw setup and temperature profile for processing of starch and cellulose nanocrystal-based nanocomposites produced by twin-screw extrusion [95].

The extrusion process can be ideally performed between 25 and 75 rpm with temperatures between 110 and 140°C. The obtained material showed good filler dispersion and improvements in stress and strain at break for filler contents smaller than 1 wt.%.

Regardless of the variations in the methodology chosen for the components addition in the extruder, it is important to note that the filler can be dispersed in starch during the gelatinization extrusion process [96].

9.6.2 Applications of Starch-Based Bio-Nanocomposites

9.6.2.1 Food industry

Food packaging and edible films are two major applications of the starch-based biodegradable polymers in food industry.

The requirements for food packaging include reducing the food losses, keeping food fresh, enhancing characteristics of food such as appearance, odour, and flavour, and provide food safety [97]. Traditional food packaging materials such as low-density polyethylene (LDPE) have the problem of environmental pollution and disposal problems [98]. The starch-based biodegradable polymers can be a possible alternative for food packaging to overcome these disadvantages while sustaining the advantages of traditional packaging materials. However, the components in the conventional starch-based polymer packaging materials are not completely static. The migration of substances into the food can possibly occur and these substances may cause harm to the human body. In view of this, new starch-based packaging materials are being developed. However, the development of a starch/clay nanocomposite food packaging material can offer better mechanical properties and reduce the migration of polymer and additives [99].

Starch-based edible films are odourless, tasteless, colourless, non-toxic, and biodegradable. They display very low permeability to oxygen at low relative humidity and are proposed for food product protection to improve quality and shelf life without impairing consumer acceptability [101]. Edible films are thin layers of edible materials applied on food products that play an important role in their preservation, distribution and marketing. The film shown in Fig. 9.12 is sourced from wheat starch and gelatin. This film can be used to control the permeability to water, oxygen and carbon dioxide, as well as lipid permeability in a food system.

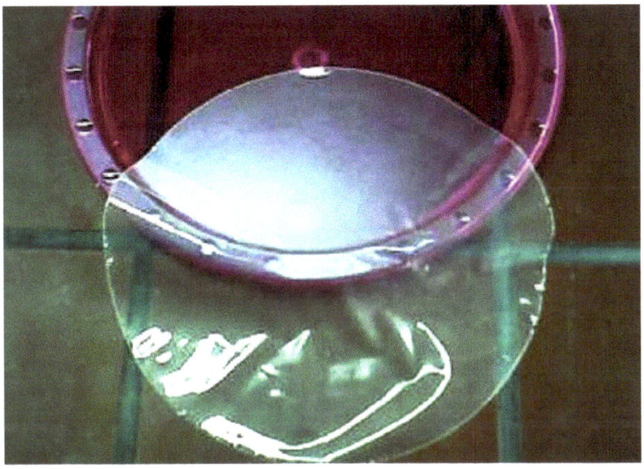

Figure 9.12 Edible wheat starch film [100].

In addition, starch can also be transformed into a foamed material by using water steam to replace the polystyrene foam as packaging material. It can be pressed into trays or disposable dishes, which are able to dissolve in water and leave a non-toxic solution [102].

Evidently, starch-based biodegradable polymers are attractive for food industry and will make great progress in the future.

9.6.2.2 Agricultural industry

Starch-based biodegradable polymers have found three major applications in agriculture: the covering of greenhouse, mulch film (as seen in Fig. 9.13), and controlled-release fertilizer materials [103]. The consumption of agriculture films is abundant. The common disposal methods of traditional films are landfill, recycling or burning. However, these methods can be time-consuming, costly and lead to environmental pollution.

On the other hand, the utilization efficiency of fertilizers is the key element of the development of agricultural productions. However, due to surface runoff, leaching and vaporization, the fertilizers escape to environment to cause diseconomy and environmental problems.

The development of starch-based biodegradable polymers offers a possibility to overcome the mentioned problems. They can be used as the fertilizers' controlled-release matrices to release the fertilizers slowly or in a controlled way. As a result, the loss of fertilizers and environment pollution can be avoided or reduced [105].

Figure 9.13 PLA mulching film [104].

After using, starch-based films can be ploughed into soil and disposed directly. Moreover, no toxic residues are formed after the degradation of starch-based biodegradable polymers. For example, to enhance the mechanical properties and solvent or gas resistance, starch-based biodegradable materials are mixed with some nano-grade additives such as TiO_2, layered silicate and MMT to form bio-nanocomposites [106].

9.6.2.3 Medical field

Starch-based biodegradable polymers have good properties to become useful medical polymer materials. These polymers possess good biocompatibility, biodegradable and its degradation products are non-toxic [107].

Starch-based biodegradable polymers have been widely investigated in bone tissue engineering.

Starch-based biodegradable bone cements can provide immediate structural support and degrade from the site of application. Moreover, they can be combined with bioactive particles, which allow new bone growth to be induced in both the interface of cement–bone and the volume left by polymer degradation. In addition, starch-based biodegradable polymer can also be used as bone tissue engineering scaffold as shown in Fig. 9.14.

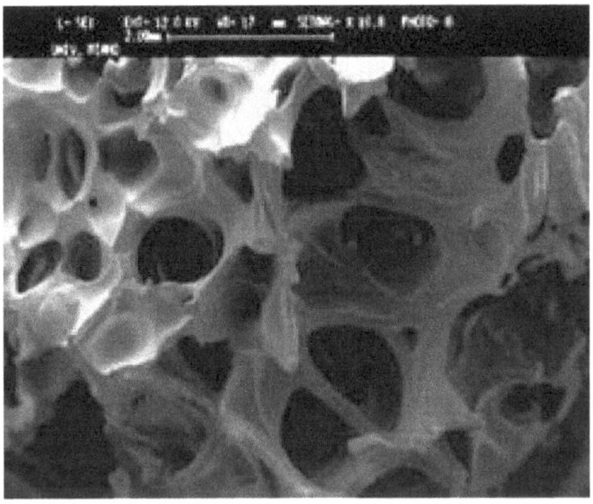

Figure 9.14 Starch-poly(lactic acid) (SPLA) scaffolds processed by using a methodology based on extrusion with blowing agents [108].

The methodology used in Fig. 9.14 is based on the mixture of the raw polymer with a blowing agent based on citric acid. After mixing, the blend undergoes extrusion or it can also be injected into a mould. During the processing, the blowing agent will degrade, releasing water and carbon dioxide which will create the pores within the polymeric matrix. By using these methods, we are able to obtain scaffolds, based on starch-based blends, with 60–70% porosity and good degrees of interconnectivity and morphologies [109].

Starch-based biodegradable polymers, in the form of a microsphere or hydrogel, are suitable for drug delivery. There is no need for surgical removal of the device after drug depletion.

The unique properties, such as hydrophilicity, permeability, biocompatibility, and make them useful for various biomedical applications [110]. The 3D structure of starch-based hydrogels enables them to absorb and reserve plenty of water and sustain their excellent mechanical properties at the same time. Starch-based hydrogels have received growing attention for biomedical applications.

9.7 Conclusion

Starch-based bio-nanocomposites are a new class of bio-based-nanomaterials, which are biodegradable and optically transparent and have higher mechanical properties, thermal stability, and barrier properties than thermoplastic starch. Starch-based bio-nanocomposites can be broadly divided into those with starch as the matrix and those with starch nanocrystals as a reinforcing phase. Cellulose nanofibers, cellulose nanocrystals, and bacterial cellulose are the most used nanoreinforcements in the starch matrix. Although these types of starch nanocomposites were usually prepared by the solution casting method, other processing methods such as twin-screw extrusion are being used to a limited extent. The studies on starch-based bio-nanocomposites have indicated that the properties of these materials vary widely based on the moisture conditions used, nature and concentrations of plasticizer and nature and concentration of the nanoreinforcement used. It is possible to obtain materials with a wide range of properties, starting from brittle material to ductile materials based on the aforementioned factors.

The potential application areas of starch-based bio-nanocomposites are in packaging, coatings, and so on. Processing methods such as twin-screw extrusion should be developed further and optimized to produce starch-based nanocomposite films and foams for packing application. In the case of medical application, the biocompatibility and nontoxicity of starch nanocomposites will be an added advantage and will find application in drug delivery, burn healing, hydrogels, and so on.

Acknowledgements

The authors thank the Technology Innovation Agency (TIA) of South Africa for funding support to carry out this project. Project reference U471.

References

1. Araújo, M. A., Cunha A., Mota, M. (2004). Enzymatic degradation of starch-based thermoplastic compounds used in prostheses: Identification of the degradation products in solution. *Biomaterials*, **25**, 2687–2693.
2. Teramoto, N., Motoyama, T., Yosomiya, R., Shibata, M. (2003). Synthesis, thermal properties, and biodegradability of propyl-etherified starch. *European Polymer Journal*, **39**, 255–261.

3. Griffin, G. J. L. (1994). Starch polymer blends. *Polymer Degradation and Stability,* **45**, 241–247.
4. Cao, X., Chen, Y., Chang, P. R., Muir, A. D., Falk, G. (2008a). Starch-based nanocomposites reinforced with flax cellulose nanocrystals. *Express Polymer Letters,* **2**, 502–510.
5. Dong, X. M., Kimura, T., Revol, J.-F., Gray, D. G. (1996). Effects of ionic strength on the isotropic-chiral nematic phase transition of suspensions of cellulose crystallites. *Langmuir,* **12**, 2076–2082.
6. Chen, Y., Liu, C., Chang, P. R., Cao, X., Anderson, D. P. (2009b). Bionanocomposites based on pea starch and cellulose Nano whiskers hydrolysed from pea hull fibre: Effect of hydrolysis time. *Carbohydrate Polymers,* **76**, 607–615.
7. Kumar, A. P., Depan, D., Singh Tomer, N., Singh, R. P. (2009). Nanoscale particles for polymer degradation and stabilization—Trends and future perspectives. *Progress in Polymer Science,* **34**, 479–415.
8. Kvien, I., Sugiyama, J., Votrubec, M., and Oksman, K. (2007). Characterization of starch based nanocomposites. *Journal of Materials Science,* **42**, 8163–8171.
9. Lu, Y., Weng, L., Cao, X. (2006). Morphological, thermal and mechanical properties of ramie crystallites—reinforced plasticized starch biocomposites. *Carbohydrate Polymers,* **63**, 198–204.
10. Angellier, H., Molina-Boisseau, S., Dole, P., Dufresne, A. (2006). Thermo plastic Starch-Waxy Maize Starch Nanocrystals Nano composites. *Biomacromolecules,* **7**, 531–539.
11. Viguié, J., Molina-Boisseau, S., Dufresne, A. (2007). Processing and characterisation of waxy maize starch films plasticised by sorbitol and reinforced with starch nanocrystals. *Macromolecular Bioscience,* **7**, 1206–1216.
12. García, N. L., Ribba, L., Dufresne, A., Aranguren, M. I., Goyanes, S. (2009). Physicomechanical properties of biodegradable starch nanocomposites. *Macromolecular Materials and Engineering,* **294**, 169–177.
13. Angellier, H., Choisnard, L., Molina-Boisseau, S., Ozil, P., Dufresne, A. (2004). Optimization of the preparation of aqueous suspensions of waxy maize starch nanocrystals using a response surface methodology. *Biomacromolecules,* **5**, 1545–1551.
14. Putaux, J.-L., Molina-Boisseau, S., Momaur, T., Dufresne, A. (2003). Platelet nanocrystals resulting from the disruption of waxy maize starch granules by acid hydrolysis. *Biomacromolecules,* **4**, 1198–1202.

15. Chang, P. R., Jian, R., Yu, J. X. (2010). MaStarch-based composites reinforced with novel chitin nanoparticles. *Carbohydrate Polymers,* **80**(2), 421–426.
16. Mathew, A. P., Dufresne, A. (2002). Morphological investigation of nanocomposites from sorbitol plasticized starch and tunicin whiskers. *Biomacromolecules,* **3**, 609–617.
17. Petersson, L., Kvien, I., Oksman, K. (2007). Structure and thermal properties of poly (lactic acid)/cellulose whiskers nanocomposite materials. *Composite Science Technology*, **67**, 2535–2544.
18. Lee, K. Y., Blaker, J. J., Bismarck, A. (2009). Surface functionalisation of bacterial cellulose as the route to produce green polylactide nanocomposites with improved properties. *Composites Science and Technology*, **69**, 2724–2733.
19. Angellier, H., Boisseau, S. M., Dole, P., Dufresne, A. (2005). Processing and structural properties of Waxy Maize starch nanocrystals reinforced natural rubber. *Macromolecules*, **38**, 3783–3792.
20. Cao, X., Chen, Y., Chang, P. R., Muir, A. D., Falk, G. (2008). Starch-based nanocomposites reinforced with flax cellulose nanocrystals. *Express Polymer Letters*, **2**, 502–510.
21. Bourtoom, T., Chinnan, M. S. (2008). Preparation and properties of rice starch-chitosan blend biodegradable film. *LWT-Food Science and Technology*, **41**, 1633–1641.
22. Petersson, L., Kvien, I., Oksman, K. (2007). Structure and thermal properties of poly (lactic acid)/cellulose whiskers nanocomposite materials. *Composite Science Technology*, **67**, 2535–2544.
23. Zimmermann, T., Pohler, E., Schwaller, P. (2005). Mechanical and morphological properties of cellulose fibril reinforced nanocomposites. *Advanced Engineering Materials,* 7, 1156–1161.
24. Paakko, M., Ankerfors, M., Kosonen, H., Nykanen, A., Ahola, S., Osterberg, M. (2007). Enzymatic hydrolysis combined with mechanical shearing and high-pressure homogenization for nanoscale cellulose fibrils and strong gels. *Biomacromolecules*, **8**(6), 1934–1941.
25. Wegner, T. H., Jones, P. E. (2006). Advancing cellulose-based nanotechnology. *Cellulose*, **13**, 115–118.
26. Samir, M. A. S. A., Alloin, F., Dufresne, A. (2005). Review of recent research into cellulosic whiskers, their properties and their application in the nanocomposite field. *Biomacromolecules*, **6**, 612–626.

27. Alemdar, A., Sain, M. (2007). Isolation and characterization of nanofibers from agricultural residues–wheat straw and soy hulls. *Bio resource Technol.*, **99**(6), 1664–1671.

28. Hulleman, S. H., Janssen, F. H. P., Feil, H. (1998). The role of water during plasticization of native starches. *Polymer*, **39**(10), 2043–2048.

29. Myllärinen, P., Partanen, R., Seppälä, J., Forssell, P. (2002). Effect of glycerol on behaviour of amylose and amylopectin films. *Carbohydrate Polymers*, **50**(4), 355–361.

30. Tomasik P., Schilling, C. H. (2004). Chemical modification of starch. *Advances in Carbohydrate Chemistry and Biochemistry*, **59**, 175–403.

31. Follain, N., Joly, C., Dole, P., Bliard, C. (2005). Mechanical properties of starch based materials. I. Short review and complementary experimental analysis. *Journal of Applied Polymer Science*, **97**, 1783–1794.

32. Lourdin, D., Coignard, L., Bizot, H., Colonna, P. (1997). Influence of equilibrium relative humidity and plasticizer concentration on the water content and glass transition of starch materials. *Polymer*, **38**(21), 5401–5406.

33. Poutanen, K, Forssell, P. (1996). *Modification of starch properties with plasticizers*, 128–131.

34. Benczédi, D, Tomka, I., Escher, F. (1998). Thermodynamics of amorphous starch-water systems. 1. Volume fluctuations. *Macromolecules*, **31**(9), 3055–3061.

35. Anglès, M. N., Dufresne, A. (2001). Plasticized starch/tunicin whiskers nano compositematerilas. 2. Mechanical behaviour. *Macromolecules*, **34**, 2921–2931.

36. Kuutti, L., Peltonen, J., Myllärinen, P., Teleman, O., Forssell, P. (1997). AFM in studies of thermoplastic starches during ageing. *Carbohydrate Polymers*, **37**(1), 7–12.

37. Anglès, M. N., Dufresne, A. (2000). Plasticized starch/tunicin whiskers nanocomposites. 1. Structural analysis. *Macromolecules*, **33**(22), 8344–8353.

38. Stading, M, Rindlav-Westling, A., Gatenholm, P. (2001). Humidity-induced transitions in amylose and amylopectin films. *Carbohydrate Polymers*, **45**(3), 209–217.

39. Wilhelm, H. M., Sierakowski, M. R., Souza, G., Wypych, F. (2003). Starch films reinforced with mineral clay. *Carbohydrate Polymers*, **52**(2), 101–110.

40. Van Soest, J. J. G., Hulleman, S. H. D., De Wit, D., Vliegenthart, J. F. G. (1996). Changes in the mechanical properties of thermoplastic potato starch in relation with changes in B-type crystallinity. *Carbohydrate Polymers*, **29**(3), 225–232.

41. Van Soest, J. J. G., De Wit, D., Vliegenthart, J. F. G. (1996). Mechanical properties of thermoplastic waxy maize starch. *Journal of Applied Polymer Science*, **61**(11), 1927–1937.

42. Forssell, P. M., Hulleman, S. H. D., Myllaärinen, P. J., Moates, G. K., Parker, R. (1999). Ageing of rubbery thermoplastic barley and oat starches. *Carbohydrate Polymers*, **39**(1), 43–51.

43. Van Soest, J. J. G., Knooren, N. (1997). Influence of glycerol and water content on the structure and properties of extruded starch plastic sheets. *Journal of Applied Polymer Science*, **64**(7), 1411–1422.

44. Dufresne, A. (2008). Polysaccharide nanocrystal reinforced nanocomposites. *Canadian Journal of Chemistry*, **86**(6), 484–494.

45. Angellier, H., Putaux, J. L., Molina-Boisseau, S., Dupeyre, D., Dufresne, A. (2005). Starch nanocrystal fillers in an acrylic polymer matrix. *Macromolecular Symposia*, **221**, 95–104.

46. Angellier, H., Molina-Boisseau, S., Lebrun, L., Dufresne, A. (2005). Processing and structural properties of waxy maize starch nanocrystals reinforced natural rubber. *Macromolecules*, **38**(9), 3783–3792.

47. Angellier, H., Molina-Boisseau, S, Belgacom, M. N., Dufresne, A. (2005). Surface chemical modification of waxy maize starch nanocrystals. *Langmuir*, **21**, 2425–2433.

48. Viguié, J., Molina-Boisseau, S., Dufresne, A. (2007). Processing and characterization of waxy maize starch films plasticized by sorbitol and reinforced with starch nanocrystals. *Macromolecular Bioscience*, **7**, 1206–1216.

49. Kristo, E., Biliaderis, C. G. (2007). Physical properties of starch nanocrystal reinforced pullulan films. *Carbohydrate Polymers*, **68**, 146–158.

50. Mele, P., Angellier-Coussy, H., Molina-Boisseau, S., Dufresne, A. (2011). Reinforcing mechanisms of starch nanocrystals in a non-vulcanized natural rubber matrix. *Biomacromolecules*, **12**, 1487–1493.

51. Chen, Y., Cao, X., Chang, P. R., Hainault, M. A. (2008). Comparative study on the films of poly (vinyl alcohol)/pea starch nanocrystals and poly (vinyl alcohol)/native pea starch. *Carbohydrate Polymers*, **73**, 8–17.

52. Dufresne, A., Vignon, M. (1998). Improvement of starch film performances using cellulose microfibrils. *Macromolecules,* **31**(8), 2693–2696.

53. Orts, W. J., Shey, J., Imam, S. H., Glenn, G. M., Guttman, M. E., Revol, J. F. (2005). Application of cellulose microfibrils in polymer nanocomposites. *Journal of Polymers and the Environment,* **13**(4), 301–306.

54. Anglès, M. N., Dufresne, A. (2001). Plasticized starch/tunicin whiskers Nano composite materials. 2. Mechanical behaviour. *Macromolecules,* **34**, 2921–2931.

55. Mathew, A. P., Thielemans, W., Dufresne, A. (2008). Mechanical properties of nanocomposites from sorbitol and tunicin whiskers. *Journal of Applied Polymer Science,* **109**, 4065–4074.

56. Teixeira, E., Pasquini, D., Curvelo, A. A. S., Corradini, E., Belgacem, M. N., Dufresne, A. (2009). Cassava bagasse cellulose nanofibrils reinforced thermoplastic cassava starch. *Carbohydrate Polymers,* **78**, 422–431.

57. Lu, Y. S., Weng, L. H., Cao, X. D. (2005). Biocomposites of plasticized starch reinforced with cellulose crystallites from cottonseed linter. *Macromolecular Bioscience*, **5**, 1101–1107.

58. Alemdar, A., Sain, M. (2008). Biocomposites from wheat straw nanofibers: Morphology, thermal and mechanical properties. *Composites Science and Technology,* **68**, 557–565.

59. Tang, X., Alavi, S., Herald, T. J. (2008a). Effects of plasticizers on the structure and properties of starch-clay nanocomposite films. *Carbohydrate Polymers,* **74,** 552–8.

60. Xu, Y., Zhou, J., Hanna, M. A. (2005). Melt-intercalated starch acetate nanocomposite foams as affected by type of organoclay. *Cereal Chemistry,* **82,** 105–10.

61. Park, H.-M., Li, X., Jin, C.-Z., Park, C.-Y., Cho, W.-J., Ha, C.-S. (2002). Preparation and properties of biodegradable thermoplastic starch/clay hybrids. *Macromolecular Materials and Engineering,* **287,** 553–8.

62. Huang, M., Yu, J., Ma, X. (2006). High mechanical performance MMT-urea and formamide-plasticized thermoplastic corn starch biodegradable nanocomposites. *Carbohydrate Polymers,* **63,** 393–9.

63. Cyras, V. P., Manfredi, L. B., Ton-That, M.-T., Vazquez, A. (2008). Physical and mechanical properties of thermoplastic starch/montmorillonite nanocomposite films. *Carbohydrate Polymers,* **73,** 55–63.

64. Wilhelm, H. M., Sierakowski, M. R., Souza, G. P., Wypych, F. (2003b). Starch films reinforced with mineral clay. *Carbohydrate Polymers*, **52**, 101–10.

65. Qiao, X., Jiang, W., Sun, K. (2005). Reinforced thermoplastic acetylated starch with layered silicates. *Starch–Stärke*, **57**, 581–6.

66. Chiou, B.-S., Yee, E., Glenn, G. M., Orts, W. J. (2005). Rheology of starch-clay nanocomposites. *Carbohydrate Polymers*, **59**, 467–75.

67. Dai, H., Chang, P., Geng, F., Yu, J., Ma, X. (2009). Preparation and properties of thermoplastic starch/montmorillonite nanocomposite using N-(2-hydroxyethyl) formamide as a new additive. *Journal of Polymers and the Environment*, **17**, 225–32.

68. Wang, N., Zhang, X., Han, N., Bai, S. (2009a). Effect of citric acid and processing on the performance of thermoplastic starch/montmorillonite nanocomposites. *Carbohydrate Polymers*, **76**, 68–73.

69. Tang, X. Z., Kumar, P., Alavi, S., Sandeep, K. P. (2012). Recent advances in biopolymers and biopolymer-based nanocomposites for food packaging materials. *Crit. Rev. Food Sci. Nutr.*, **52**, 426–442.

70. Zeng, S., Wu, X., Lin, S., Zeng, H., Lu, X., Zhang, Y., Zheng, B. (2015) Structural characteristics and physicochemical properties of lotus seed resistant starch prepared by different methods. *Food Chem.*, **186**, 213–222.

71. Kugge, C., Johnson, B. (2008). Improved barrier properties of double dispersion coated liner. *Prog. Org. Coat.*, **62**, 430–435.

72. Dufresne, A. (2013). Nanocellulose: A new ageless bionanomaterial. *Mater. Today*, **16**, 220–227.

73. Narayan, R., Pettigrew, C. (1999). ASTM standards help define and grow a new biodegradable plastics industry. ASTM Standardization News. December: 36–42.

74. Hung, P. V., Vien, N. L., Lan Phi, N. T. (2016). Resistant starch improvement of rice starches under a combination of acid and heat-moisture treatments. *Food Chem.*, **191**, 67–73.

75. Huang, T. T., Zhou, D. N., Jin, Z. Y., Xu, X. M., Chen, H. Q. (2015). Effect of debranching and heat-moisture treatments on structural characteristics and digestibility of sweet potato starch. *Food Chem.* **187**, 218–224.

76. Martin, O., Schwach, E., Averous, L. Couturier, Y. (2001). Properties of biodegradable multilayer films based on plasticized wheat starch. *Starch.*, **53**(8), 372–380.

77. Bismarck, A., Aranberri-Askargorta, I., Springer, J., Lampke, T., Wielage, B., Samboulis, A., Shenderovick, I., Limbach, H. (2002). Surface characterization of flax, hemp, and cellulose fibers; Surface properties and the water uptake behavior. *Polymer Composites*. **23**(5), 872–894.

78. Hornsby, P., Hinrichsen, E., Tarverdi, K. (1997). Mechanical properties of polypropylene composites reinforced with wheat and flax straw fibers. *Journal of Materials Science*, **32**(4), 1009–1015.

79. Chandra, R., Rustgi, R. (1998). Biodegradable polymers. *Progressive Polymer Science*, **23**, 1273–1335.

80. Frisoni, G., Baiardo, M., Scandola, M. (2001). Natural cellulose fibers: Heterogeneous acetylation kinetics and biodegradation behavior. *Biomacromolecules*, **2**, 476–482.

81. Bledzki, A. K., Gassan, J. (1999). Composites reinforced with cellulose based fibers. *Progress in Polymer Science*, **24**, 221–274.

82. Chiellini, E., Cinelli, P., Imam, S., Mao, L. (2001). Composite films based on biorelated agroindustrial waste and PVA. Preparation and mechanical properties characterization. *Biomacromolecules*, **2**, 1029–1037.

83. Angles, M. N., Dufresne, A. (2001). Plasticized starch/tunicin whiskers nanocomposites. 2. Mechanical behaviour. *Macromolecules*, **34**, 2921–2931.

84. Mathew, A. P., Chakraborty, A., Oksman, K., Sain, M. (2006). The structure and mechanical properties of cellulose nanocomposites prepared by twin screw extrusion. *ACS Symp. Series*, **938**, 114–131.

85. Kaushik, A., Mandeep, S. (2011). Isolation and characterization of cellulose nanofibrils from wheat straw using steam explosion coupled with high shear homogenization. *Carbohydr. Res.*, **346**, 76–85.

86. Lu, Y., Weng, L., Cao, X. (2006). Morphological thermal and mechanical properties of ramie crystallites-reinforced plasticized starch biocomposites. *Carbohydrate. Polymers.*, **63**, 198–204.

87. Angellier, H., Boisseau, S. M., Lebrun, L., Dufresne, A. (2005). Processing and structural properties of waxy maize starch nanocrystals reinforced natural rubber. *Macromolecules*, **38**, 3783–3792.

88. Deeborah, L. C., Julien, B., Dufresne, A. (2010). Starch nanoparticles: a review. *Biomacromolecules*, **11**, 1153–1162.

89. Funke, U., W. Bergthaller, Lindhauer, M. G. (1998). Processing and characterization of biodegradable products based on starch. *Polym. Degrad. Stabil.*, **59**(1–3), 293–296.

90. Zheng, H., Ai, F., Chang, P. R., Huang, J., Dufresne, A. (2009). Structure and properties of starch nanocrystal-reinforced soy protein plastics. *Polym. Compos.*, **30**(4), 474–480.
91. Tadmor, Zehev Costas G. Gogos, (2006). Principles of Polymer Processing. Joboken, New Jersey: John Wiley & Sons.
92. Osswald Tim A. (1998). Polymer Processing Fundamentals. Cincinnati: Gardner Publications.
93. Kutz, Myer. (2011). Applied Plastics Engineering Handbook: Processing and Materials. Elsevier Inc.
94. Joshi, M., Aldred, P., McKnight, S., Panozzo, J. F., Kasapis, S., Adhikari, R., Adhikari, B. (2013). Physicochemical and functional characteristics of lentil starch. *Carbohydr. Polym.*, **92**, 1484–1496.
95. Klemens, K., Werner, W. (2008). Co-rotating twin-screw extruders, fundamentals. *Technol. Appl.*, **44**.
96. Kalpakjian, Serope Steven Schmid, R. (2010). *Manufacturing Engineering and Technology* 6th Ed. Pearson Education, Inc.
97. Ozdemir, M., Floros, J. D. (2004). Active food packaging technologies. *Critical Reviews in Food Science and Nutrition*, **44**, 185–193.
98. Avella, M., de Vlieger, J. J., Errico, M. E., Fischer, S., Vacca, P., Volpe, M. G. (2005). Biodegradable starch/clay nanocomposite films for food packaging applications. *Food Chemistry*, **93**, 467–474.
99. The, D. P., Debeaufort, F., Voilley, A., Luu, D. (2009). Biopolymer interactions affect the functional properties of edible films based on agar, cassava starch and arabinoxylan blends. *Journal of Food Engineering*, **90**, 548–558.
100. Flores, S., Haedo, A. S., Campos, C., Gerschenson, L. (2007). Antimicrobial performance of potassium sorbate supported in tapioca starch edible films. *European Food Research Technology*, **225**, 375–384.
101. Siracusa, V., Rocculi, P., Romani, S., Rosa, M. D. (2008). Biodegradable polymers for food packaging: A review. *Trends in Food Science and Technology*, **19**, 634–643.
102. Dilara, P. A., Briassoulis, D. (2000). Degradation and stabilization of low-density polyethylene films used as greenhouse covering materials. *Journal of Agricultural Engineering Research*, **76**, 309–321.
103. Bohlmann, G., Toki, G. (2004). *Chemical economics handbook*. SRI International, Menlo Park.

104. Dave, A. M., Mehta, M. H., Aminabhavi, T. M., Kulkarni, A. R., Soppimath K. S. (1999). A review on controlled release of nitrogen fertilizers through polymeric membrane devices. *Polymer-Plastics Technology and Engineering*, **38**, 675–711.

105. Guo, M., Liu, M., Zhan, F., Wu, L. (2005). Preparation and properties of a slow-release membrane-encapsulated urea fertilizer with superabsorbent and moisture preservation. *Industrial and Engineering Chemistry Research*, **44**, 4206–4211.

106. Kumbar, S. G., Kulkarni, A. R., Dave, A. M., Aminabha, T. M. (2001). Encapsulation efficiency and release kinetics of solid and liquid pesticides through urea formaldehyde crosslinked starch, guar gum, and starch + guar gum matrices. *Journal of Applied Polymer Science*, **82**, 2863–2866.

107. Reddy, S. M., Sinha, V. R., Reddy, D. S. (1999). Novel oral colon-specific drug delivery systems for pharmacotherapy of peptides and nonpeptide drugs. *Drugs of Today*, **35**, 537–580.

108. Sinha, V. R., Kumria, R. (2001). Polysaccharides in colon-specific drug delivery. *International Journal of Pharmaceutics*, **224**, 19–38.

109. Sun, Q., Dai, L., Nan, C., Xiong, L. (2014). Effect of heat moisture treatment on physicochemical and morphological properties of wheat starch and xylitol mixture. *Food Chem.*, **143**, 54–59.

110. Boesel, L. F., Mano, J. F., Reis, R. L. (2004). Optimization of the formulation and mechanical properties of starch based partially degradable bone cements. *Journal of Materials Science: Materials in Medicine*, **15**, 73–83.

Chapter 10

Eggshells: From Waste to Medical Applications

Stanley Chibuzor Onwubu,[a] Shalini Singh,[b] Anisa Vahed,[a] and Krishnan Kanny[c]

[a]*Department of Dental Sciences, Durban University of Technology, South Africa*
[b]*Department of Operations and Quality, Durban University of Technology, South Africa*
[c]*Department of Mechanical Engineering, Durban University of Technology, South Africa*

shalinis@dut.ac.za

The use of liquid eggs at the industrial level has resulted in a considerable amount of eggshell waste as a by-product. Most of the residual waste eggshells from these industries are mainly discarded into landfills without further treatment, subsequently overburdening the environment. Given the amount of eggshell waste that needs to be disposed of worldwide, there is a need to find alternative means of converting eggshell waste materials into value-added products for environmental sustainability. Eggshell, together with its membrane, is also rich in minerals and proteins such as calcium carbonates, glycoprotein, proteoglycan, and collagen. Despite the unique chemical composition of eggshell, it has not extensively gained attention for its medical benefits.

Handbook of Biopolymers: Advances and Multifaceted Applications
Edited by Shakeel Ahmed, Suvardhan Kanchi, and Gopalakrishnan Kumar

ISBN 978-981-4800-17-4 (Hardcover), 978-0-429-02475-7 (eBook)
www.panstanford.com

Eggshell is an inexpensive and abundant biomaterial with interesting properties, which can be tailored to numerous potential applications. This chapter foregrounds the promising area of research in the application of eggshell waste for medical and dental benefits. Eggshells have the potential of synthesis into hydroxyapatite, which is a main component of bones and teeth. Eggshell-derived hydroxyapatite reportedly showed excellent biocompatibility with soft tissues such as skin, muscle and gums, making eggshell waste an ideal biomaterial for bone repair, tissue regeneration and dental implants. Eggshell powder has a calcium content of approximately 38%, which offers a promising but little utilized source of calcium for human supplement. The membrane within the eggshell is highly prized for pharmaceutical usage. Eggshell membrane consists of collagen as a component. Collagen is a type of protein that connects and supports other bodily tissues, such as skin, bone, tendons, muscles and cartilage. Eggshell membrane collagen has very low autoimmune and allergic reactions as well as high biosafety in humans. Eggshell and its associated membrane have also shown excellent results in cosmetics, as exfoliating agents in beauty treatment, abrasive materials in smoothening of dentures and expediency in drug delivery. Using eggshell waste in medical and dental applications provides an opportunity to reduce the cost of medical and dental treatment with minimal impact on the environment.

10.1 Introduction

In the past decade, much attention has been given to biodegradable therapeutic materials because of their outstanding healing abilities. Both synthetic and naturally derived biodegradable polymeric materials have been investigated extensively as biomaterials for bone tissue regeneration and construction. Among these materials, hydroxyapatite polymer composite has been used as a biocomposite and osteo-conductive substitute for bone repair and implantation. This chapter highlights the use of eggshell-derived hydroxyapatite as an adaptive innovative bone graft substitute that has shown favourable results in the treatment of bone and dental defects. This was attributed to its biocompatibility with polymeric materials, complete bone regeneration, lack of disease transfer, ease of use, and unlimited availability. The chapter also discusses the medical application of collagen extracted from eggshell membrane.

Eggshell is a natural bioceramic composite that has a unique chemical makeup of inorganic and organic compounds. It consists of an inorganic shell and an organic membrane. The shell together with its membrane weighs approximately 11% of the egg weight. The inorganic constituent of eggshells is mainly calcium carbonates (~ 95% in the form of calcite), whereas its organic compounds are a matrix of proteins, glycoprotein, proteoglycan, and type X collagen sulphated polysaccharides (~ 5%), among others. Currently, the global estimate for egg production is 65.5 million metric tons per year with Asia leading the world output. In South Africa, the total production of eggs is estimated to be 133.9 million per week [1]. According to Cristianne et al. [2], a large percentage of eggs produced worldwide are diverted to breaker operations to produce liquid egg products, where eggshells are not consumed. For example, 32% of eggs produced in the United States and approximately 25% in Europe are diverted for the production of liquid eggs.

10.1.1 Problems Arising from Waste Eggshells

Most of the eggshell waste produced from these industries are simply utilized as fertilizers or discarded into landfills. Das et al. [5] also pointed out that industries incur high costs in the disposal of eggshells. Choi and Lee [3] argued that many landfills in most countries are unwilling to take more eggshell waste because it attracts vermin. Similarly, Bitzer and Sims [4] reported that eggshells used for cropping causes nitrate contamination of ground water. This is concerning as there is a strong correlation between the level of nitrate in drinking water and the occurrence of blue baby syndrome, cancer, respiratory illnesses in humans and foetal abortions in livestock. Given the amount of eggshell waste that needs to be disposed worldwide, it is important to find alternative means of converting eggshell waste materials into value added products for environment sustainability.

10.1.2 Motivation for the Application of Eggshells

There has been an increase in research and development in processes to extract products with potential commercial applications from eggshell waste. In the past decade, eggshells have reportedly been used as an additive for animal feed and human nutrition, coating pigment for inkjet printing, dye-effluents remover, and as bio-fillers for polymer-based composites. Recently, there has been increased interest in the use of eggshells and its associated membrane for a variety of medical applications. For example, Abdulrahman et al. [6] demonstrated how calcium carbonate derived from eggshells could be used to produce hydroxyapatite, which is mainly used in bone and dental treatments. They argued that eggshells based hydroxyapatite and nanohydroxyapatite could reduce the cost of treatment in bone repair or replacement.

Moreover, and as noted by Oliveira et al. [7], the use of eggshells as an alternative source of calcium carbonate may alleviate the impact on the natural reserves of limestone, which is a non-renewable natural source of calcium carbonate [8]. A noteworthy point is that calcium carbonate is a rich source of mineral for the pharmaceutical industry and is a base material for developing medicinal and dental preparations. Murakami et al. [9] noted that calcium carbonate from eggshells has great potential in the development of alternative ingredients for medical applications such as food additive and calcium supplement, and as a component for bone implants. They also elaborated that the dearth of information on the medical application of eggshells suggests that there is a promising area worth exploring. Furthermore, Oliveira et al. [7] indicated that glycoproteins, which consist of Type I, V and X collagen are useful in the production of cosmetics. Keratin, dermatan sulphate and glycosaminoglycans other constituents of eggshells are used in the production of collagen and for the synthesis of other products such as chitin. Similarly, Fatima et al. [10] reported that sialic acid, a sial-oligosaccharide, a constituent in eggshells has therapeutic properties for infectious diseases such as influenza. Another constituent in the eggshell membrane, bacteriolytic enzymes such as lysozyme and beta-N-acetyl glucosaminidase, alters the thermal resistance of bacteria. Lastly, amino acids such as lysine, proline, alanine, cysteine and phenylalanine contained in eggshells are used for medical applications. Despite the unique chemical composition of eggshell and its potential use in medical applications, it has not extensively gained attention for its medical benefits.

This chapter therefore aims to give an insight into the current medical and dental applications of eggshells. The chapter will first provide an overview of literature related to eggshells. The focus will be on the extraction of the starting material from eggshell powder for the synthesis of hydroxyapatite. According to Dutta et al. [11], hydroxyapatite (Fig. 10.1) is the mineral component found in bones (60–70%) and dental enamel (98%) with a chemical composition of $Ca_{10}(PO_4)_6(OH)_2$. It is therefore reasonable to assume that hydroxyapatite derived from eggshells could be used to rehabilitate and regenerate bone and teeth. The benefits and medical importance of hydroxyapatite will be detailed in the next section of this chapter. The use of eggshells as a calcium supplement in general and eggshell membrane in the production of collagen will be subsequently discussed.

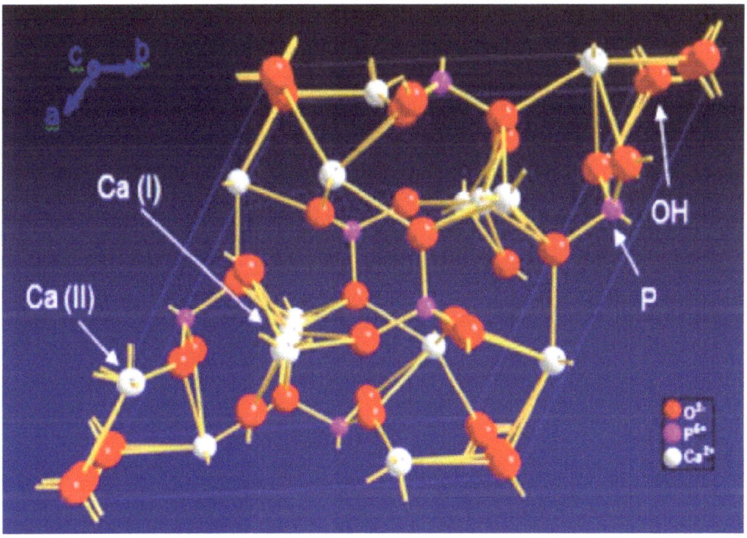

Figure 10.1 Structure of hydroxyapatite. *Source*: [11].

10.2 Overview of Eggshells Structure and Properties

Generally, and as highlighted by several studies [12–14], eggshells are lightweight, crushable, texturally hard and porous to protect the chicken embryo from microorganisms. Freire and Holanda [15] emphasized that the porousness of eggshells enables it to be crushed into various particle sizes ranging from 2 to 900 microns. Consequently, the eggshell powder produced could potentially be used for various applications.

Hincke et al. [16] revealed that the avian eggshells epitomizes the most advanced amniotic egg in oviparous vertebrates. Importantly, avian eggshells function to regulate the exchange of metabolic gases and water while protecting the contents of the egg from microbes and the physical environment. Hincke et al. [16] stated that many physiological, biochemical, structural and morphological studies conducted on avian eggs used domestic chicken eggs or *Gallus gallus*. Unlike other avian shells such as ostrich, duck, goose, and elephant bird, chicken eggs are abundantly available and renewable [16]. Hence, Section 10.2 highlights the considerable body of research dealing with the domestic chicken eggshells, which has provided insight into its structure and properties.

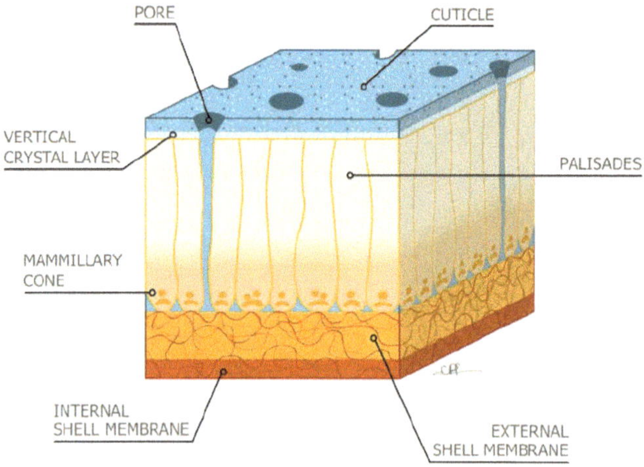

Figure 10.2 Eggshells structure. *Source*: [16].

As illustrated in Fig. 10.2, Chien et al. [17] and Kang et al. [18] suggest that the chicken eggshell has a well-defined structure consisting of the mammillary body layer, thick palisade layer, and transitional vertical crystal layer. A non-calcified cuticle layer coats the eggshells. The earlier work of Dennis et al. [19] noted that the transitional, inner zone of the eggshells cuticle contains spherical aggregates of hydroxyapatite. Owing to the unique structure of eggshells, attempts have been made to synthesize hydroxyapatite from chicken eggshells in commercial quantity using different methods of extraction. The next section will therefore examine the progress made in the synthesis of hydroxyapatite from eggshells.

10.3 Synthesis of Hydroxyapatite from Eggshells

Generally, synthetic hydroxyapatite is commonly used in biomaterial engineering and regenerative medicine to speed up bone repair [20]. Lee et al. [21] compared the regenerative ability of synthetic hydroxyapatite against those derived naturally from eggshells in the healing of bone defect in a rabbit. They observed that greater bone formation was achieved with the hydroxyapatite derived from eggshells. Consequently, researchers are directing their attention to the use of different natural calcium carbonate base materials such as seashells, bovine bone, cuttlefish, oyster shells, and eggshells in the synthesis of hydroxyapatite for biomedical applications. Essentially, Prabakaran et al. [22], Ho et al. [23], and Kamalanathan et al. [24] noted that the high calcium carbonate content, together with the trace amounts of ions such as Na^{+}, Sr^{2+}, and Mg^{2+}, makes eggshells a useful waste material for the synthesis of hydroxyapatite. Akram et al. [25] emphasized that the presence of the aforementioned ions in eggshells directly influences the biochemical reactions associated with bone metabolism. Unlike eggshells that are readily available, Wu et al. [26] cautioned that the variable nature of other biological sources of hydroxyapatite limits its use in hydroxyapatite production.

From an environmental sustainability perspective, Abdulrahman et al. [6] advised that using eggshells to synthesize hydroxyapatite would significantly reduce the pollution effects of waste materials and increase the production of highly valuable products. They further stated that the process of the conversion of eggshells into hydroxyapatite and nanohydroxypatite is an environmental friendly process. As such, hydroxyapatite derived from eggshells could potentially reduce the cost of treatment in bone repair and replacement with little impact on the environment [27]. It is worth mentioning that different methods and approaches such as hydrothermal method (Fig. 10.3) [28–30], wet chemical method (Fig. 10.4) [27, 31], microwave irradiation (Fig. 10.5) [32], sintering [25] and high-energy mechanical activation [33, 34] have been proposed in the literature for hydroxyapatite production from eggshells. Despite the different methods used in synthesizing hydroxyapatite from eggshells, Abdulrahman et al. [6], suggested that hydroxyapatite produced from any of these methods would have excellent biocompatibility and mechanical strength thereby facilitating effective tissue repair and regeneration. Table 10.1 illustrates the different characteristics and properties of hydroxyapatite produced using different methods of synthesizing hydroxyapatite and is the point of departure for the next section.

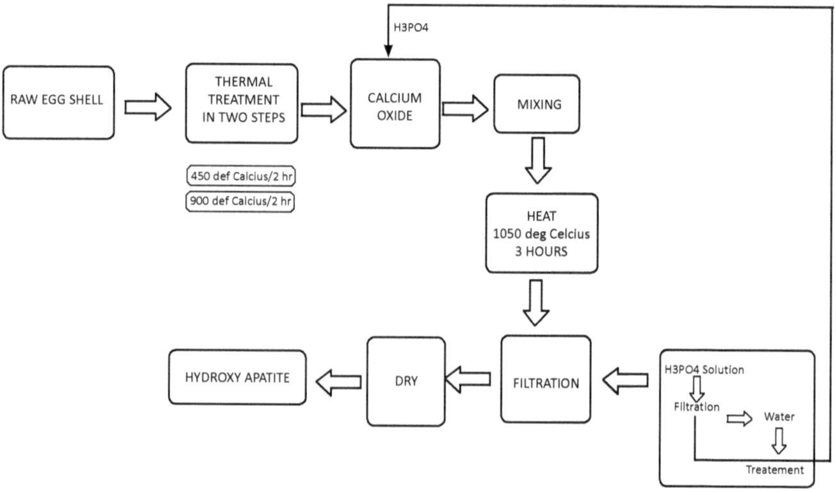

Figure 10.3 Hydroxyapatite process flowchart. Adapted from [7].

Table 10.1 Characteristic and properties of hydroxyapatite synthesis from eggshells using different methods

Methods of synthesizing	Starting material	Requirement	Unique properties of the hydroxyapatite produced	Characterization techniques
Hydrothermal	$Ca(OH)_2$ derived from eggshells and $CaHPO_{4.2}H_20$	High temperature following three temperature range: 450°C for the first stage, between 450–600°C for the second stage, and finally between 600–950°C.	Fine hydroxyapatite single crystal Excellent homogeneity and high crystallinity	1. The phase identification and characterization of the synthesized hydroxyapatite is usually confirmed by
Microwave irradiation	Ca derived from eggshells	Chelating agent such as ethylene diamine tetra acetic acid (EDTA)	High sinterability and stability at high temperature Good morphological properties.	(a) X-ray diffraction (b) Fourier Transform infrared spectroscopy.
Mechanochemical activation process	Eggshells	Ball-milling or attrition milling for 20 h	Single phase hydroxyapatite of high crystallinity, homogenous hydroxyapatite with a spherical like or rod-like morphology.	2. The particle size and morphology of the synthesized hydroxyapatite are identified using
Wet chemical method	Calcium oxide derived from eggshells and phosphoric acid	Conversion of eggshells to calcium oxide in a furnace at 900°C for 1 h to Thereafter, 0.6 M solution of orthophosphoric acid is added to calcined eggshells.	Highly pure nanocrystalline hydroxyapatite powder with an irregular morphology.	(a) Scanning electron microscope. (b) Transmission electron microscope
Sintering method	Calcined eggshell powder and dicalcium hydrogen phosphate di-hydrate	Heat treatment At 800 followed by sintering temperature ranging from 1050 to 1350 for 2 h	Superior sinterability and osteoconduction, phase pure highly crystalline hydroxyapatite with a flower-like morphology.	(c) Light scattering analysis.

Source: Abdulrahman et al. [6].

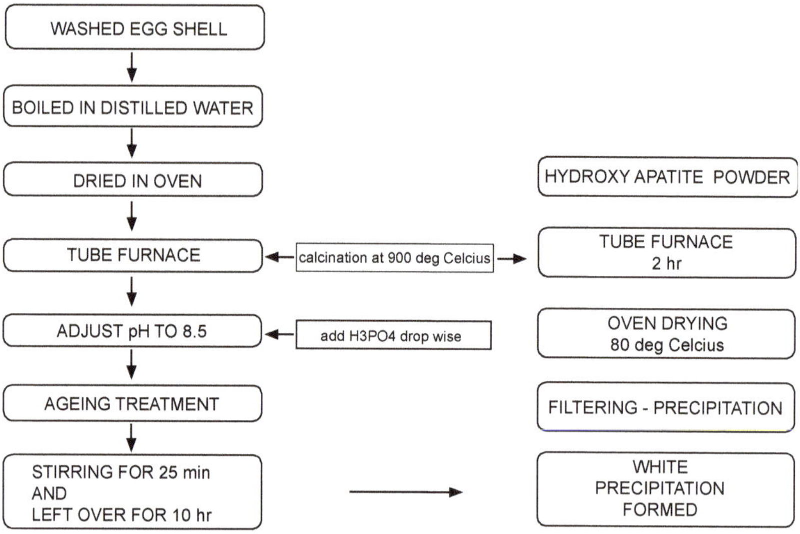

Figure 10.4 Schematic process flow chart for synthesis of hydroxyapatite powder by wet chemical method. Adapted from [31].

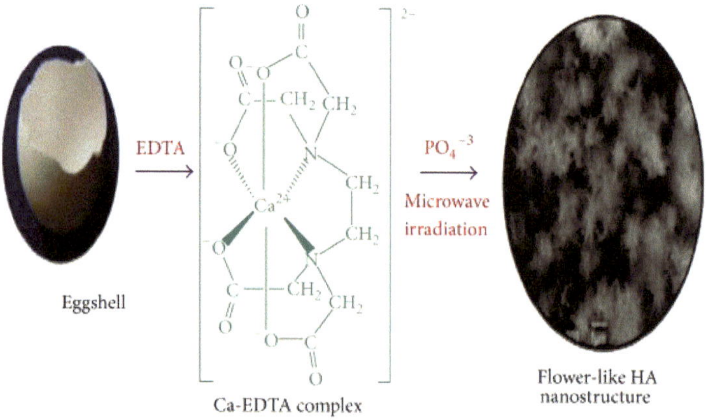

Figure 10.5 Microwave conversion of eggshells into a flower-like hydroxyapatite. *Source*: [32].

10.4 Medical Benefit of Eggshell-Derived Hydroxyapatite

According to Akram et al. [25] and Ramesh et al. [35], the healing of segmented bone defects is a challenge in orthopaedic and trauma surgery because of the limited availability of bone material needed to fill the defect and promote bone growth. In attempting to improve the management and treatment of bone defects, bone graft substitutes such as hydroxyapatite, demineralized allograft bone matrix and bovine collagen mineral composites are being considered. Importantly, and as noted by Oliveira et al. [7] and Khandelwal and Prakash [31], hydroxyapatite's mineral composition is similar to bone and is biocompatible with soft tissues and surrounding anatomical structures. Kamalanathan et al. [36] also pointed out that the advantage of hydroxyapatite obtained from biological sources like eggshells stems from its similarity to human bone (Table 10.2). These positive attributes have made hydroxyapatite the common material used for bone grafts, orthopaedic and dental implants.

Table 10.2 Comparison of hydroxyapatite synthesis from eggshells and other materials

Parameters	Bovine bone	Fish bones	Eggshells	Natural bone
Ca/P ratio	>1.67	≥1.67	≥1.67	≥1.67
Particle size	Micro to nanometre range	In nanometre range	≥50 nm	Nanocomposite
Morphology	Needle, rod, plate, spherical	Needle, rod, spherical	Needle or rod	Needle or rod
Phase composition	Phase pure and tricalcium phosphate	Phase pure	Phase pure	Composite
Thermal stability	800–1000°C	600–800	<800	>1200
Trace elements	Na^+, Mg^{2+}, K^+, Ti^{2+}	Fe^{3+}, Cr^{3+}, Cu^{2+}, K^+, Mg^{2+}	CO_3^{2-} major element, Mg^{2+}, Na^+	Na^+, Mg^{2+}, K^+, $Zn^2 + CO_3^{2-}$

Source: [6, 25].

Macha et al. [37] outlined that eggshells have been used in biomedical applications directly as a bone substitute either for maxillofacial surgery or as sources of hydroxyapatite in bioceramic production. They noted that the progress recently achieved in orthopaedic and dental applications have increased the demand of eggshells natural hydroxyapatite bioceramics.

For example, Kattimani et al. [38] reported that eggshell-derived hydroxyapatites showed early bone regeneration and values of density measurements equal or more than that of the surrounding normal bone. Their work supports other studies [21, 39] were eggshells-derived hydroxyapatite was highly effective in the treatment of bone defects in rabbit graft models. Similarly, Kattimani et al. [40] reported that eggshell-derived hydroxyapatite showed complete bone formation in the treatment of human maxillary cystic bone defects. The aforementioned studies therefore provide evidence that eggshell-derived hydroxyapatite is the gold standard in terms of bone regeneration.

10.5 Eggshells as a Source of Calcium Supplement

Broadus [41] revealed that calcium is an essential requirement to the human body for a range of physiological functions and for the maintenance of bone tissues throughout life. Resonating with Broadus [41], Gaonkar and Chakraborty [42] asserted that in the human body 99% of calcium is stored in bones and teeth. The other 1% is found in blood to support critical functions such as controlling the body blood pressure and the maintenance of heartbeat.

Brun et al. [43] and Hassan [44], however, have argued that the consumption of calcium is below the established clinical guidelines (Table 10.3) leading to calcium deficiency. Consistent with the aforementioned authors, Gaonkar and Chakraborty [42] asserted that when the body calcium intake becomes too low to keep the blood calcium normal, the body withdraws the calcium it needs from the bones. Over time, and when the amount of calcium taken out from the bones is higher than those deposited in the bone, calcium will be depleted from the bone bank. This causes the bones to weaken and to be susceptible to fracture and breakage. To resolve this problem, calcium supplements are recommended. Hasan [44] however, contended that the high costs associated with calcium supplements make levels of adherence difficult, particularly in developing countries such Africa, parts of Asia and the Caribbean were poverty is high.

This underpinning factor therefore supports the use of eggshells as a viable option in calcium supplements.

Table 10.3 The amount of calcium required by the human body

Age	Calcium needed each day (mg)*
Birth to 12 months	Supplied by formula or breast milk
1–3 years	700
4–8 years	1000
9–18 years	1300
Men 19–70 Women 19–50	1000
Women 51–70	1200

Source: Gaonkar and Chakraborty [42].

In addition, and as revealed by Hassan [44], chicken eggshell powder has a calcium content of approximately 38%, which offers a promising but little utilized source of calcium for human supplement. Notwithstanding this, he further elaborated that calcium derived from eggshells is 90% absorbable in the human body. This aligns with the early work of Oguido et al. [45], who noted that the absorption of calcium from eggshells is greater in males when compared with the absorption of calcium found in milk. Oguido et al. highlight the potential benefits of eggshell powder in the treatment of bone disease associated with calcium deficiency.

In terms of performance equivalency, Gaonkar and Chakraborty [42] pointed out that eggshell powder is comparable against CIPCAL-500 calcium supplement. A critical point worth mentioning is that a number of animal and human studies carried out primarily in Europe and Asia found eggshells to be completely safe for both human consumption and animal feeds [46, 48]. Essentially, the Association of American Feed Control Officials legitimately recognized eggshells meal [49]. As summarized in Table 10.4, Hassan [40] confirmed eggshell powder is a suitable and economical source of calcium for human supplements.

Table 10.4 Chemical composition of eggshell powder and wheat

Parameter (%)	Wheat flour	Eggshell powder
Crude protein	9.50	5.40
Fat	0.80	0.02
Ash	0.45	90.2
Calcium	49.95 mg/100 g	35080 mg/100 g
Phosphorus	121.2 mg/100 g	150.2 mg/100 g
Magnesium	30.2 mg/100 g	262.0 mg/100 g
Iron	1.05 mg/100 g	13.06 mg/100 g
Zinc	0.55 mg/100 g	145.1 mg/100 g
Potassium	111.21 mg/100 g	50.00 mg/100 g
Sodium	55.9 mg/100 g	47.9 mg/100 g
Copper	0.20 mg/100 g	4.1 mg/100 g
Manganese	0.25 mg/100 g	149.9 mg/100 g

Source: Hassan [44].

It is worth noting that a single eggshell has a mass of 6 g, which can yield around 2200 mg calcium [50], Bee [51] and King'Ori [52] reported that one medium sized eggshell produces approximately 750–800 mg of calcium. They advised that eggshells used for calcium be boiled in hot water for 5–10 minutes to kill pathogens and subsequently dried and ground into fine powder. Suzuki et al. [53] patented another method of producing calcium from eggshells, which is free of pathogen. This entails thermal treatment of the eggshells at three different stages. As illustrated in Fig. 10.6, the calcining temperature used ranges from 350°C (3a), to 600°C (4a) and 1200°C (5a). The eggshells developed through these stages are free of pathogens and have antimicrobial properties, which is significant if used as calcium supplements for humans.

From a medical context, it is recommended that people suffering from bone defects such as osteoporosis require 400–500 mg calcium per day to supplement dietary sources [51]. According to Hassan [44], the use of eggshell powder as calcium supplement could increase bone density and reduce pain in patients with osteoporosis. Similarly, a clinical study by Schaafsma and Pakan [47] revealed osteoporotic postmenopausal women who received eggshell powder together with vitamin D3 and magnesium supplement their lumbar spine, total proximal femur and trochanter bone mineral density increased. The early work of Makai and Chudacek [54] also showed that eggshell powder increased bone mineral density and decreased pain among elderly populations. Consistent with these studies, Hirasawa et al. [48] reported that the bone mineral density of the lumbar spine and the proximal tibia improved in rats treated with eggshell powder.

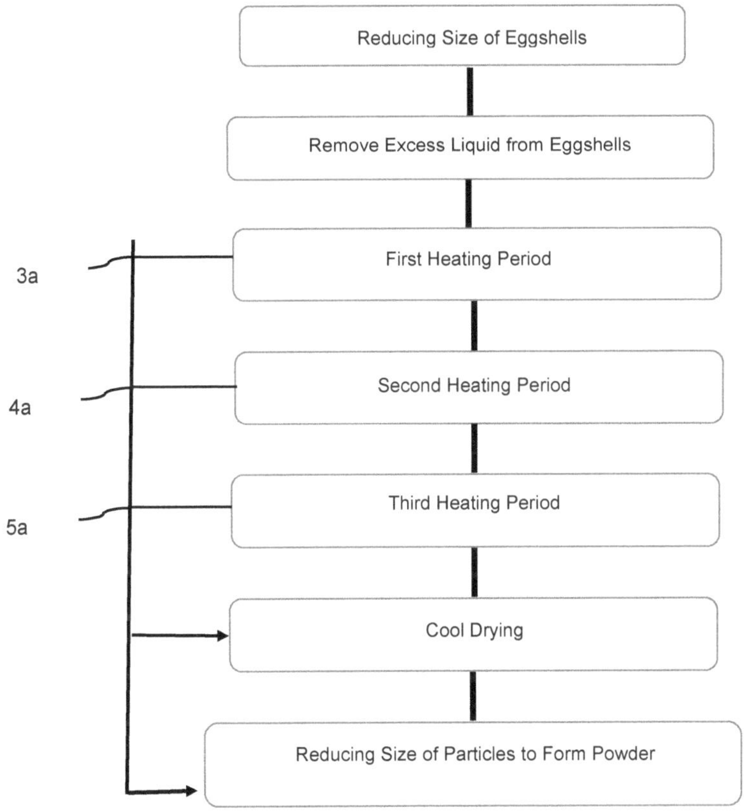

Figure 10.6 Method of producing eggshell powder free of pathogens. Adapted from [53].

10.6 Medical Application of Collagen Extracted from Eggshell Membrane

For pharmaceutical and cosmetic industries, eggs are much more than yolks and whites. According to the European Commission for research and innovation [55], the collagen found within the eggshell membrane is highly prized for pharmaceutical usage. Rose-Martel and Hincke [56] reported that collagen represents approximately 10% of the eggshell membrane fibre protein. Zhao and Chi [57] pointed out that the collagen in eggshell membrane and that of mammalian share similar characteristics. King'Ori [52] explained that collagen is a type of protein that connects and supports other bodily tissues such as skin; bone tendons; muscles and cartilage, teeth; and internal organs. In terms of collagen functions and properties in humans, Madison [58] argued that collagen works in conjunction with elastin in supporting the body's tissues. He elaborated that collagen gives body tissues form and provides firmness and strength, whereas elastin provides the same body tissues much need flexibility. Madison also emphasized that the combination of collagen and elastin is important in many parts of the body such as the lungs, bones, tendons, and blood vessels. Collagen is widely used in cosmetic surgery as a healing aid for burn patients, in the reconstruction of bone and for various dental and orthopaedic purposes. In the past, collagen use for medical applications has been isolated mainly from bovine, swine skins and bones. King' Ori [52] reported that the outbreaks of bovine spongioform encephalopathy (mad cow disease), foot and mouth disease, autoimmune and allergic reactions has restricted the use of collagen from this source. In contrast, Long et al. [59] noted that eggshell membrane collagen is low in autoimmune and allergic reactions, while being high in bio-safety. Consequently, safer alternative sources for collagen extraction such as eggshell membrane are being researched.

Collagen is often extracted using one of three methods, namely (1) organic acid such as acetic and citric acid, (2) organic amine such as methylenediamine hydrochloride and (3) acid-pepsin digestion. Ponkham et al. [60] reported on the extraction of collagen from eggshell membrane using organic acids such as acetic acid and citric acid. They revealed yields of 507 and 495 mg for

Type I collagen with 0.5 M acetic and 0.5 M citric acid, respectively. Similarly, Zhao and Chi [57] reported on the extraction of collagen from eggshell membrane using pepsin digestion (Fig. 10.7). The collagen extracted using pepsin digestion can be typified as Type I collagen. It was noted that the three kinds of collagen extracted from eggshell membrane can be used in various fields [52]. Equally important is that 120,000 tons of eggshell waste would yield 110,000 tons of eggshells and 10,000 tons of eggshell membrane. Bearing in mind that 10% of membrane is collagen, it would yield 1,000 pounds of collagen, which is currently priced at $ 1000 per gram or about 454,000 per pound [61].

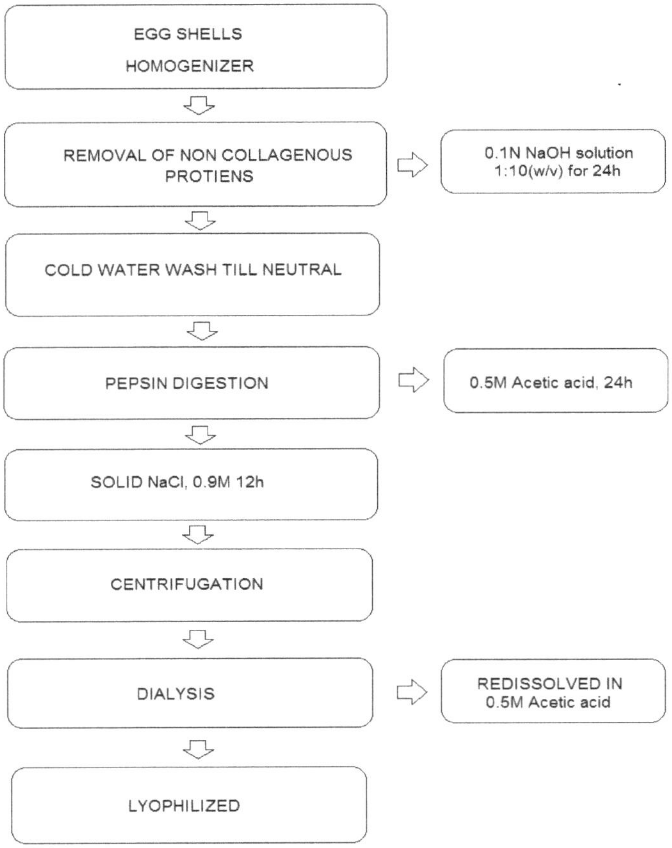

Figure 10.7 Schematic illustration showing pepsin digestion process of producing collagen from eggshell membrane. Adapted from [57].

Drawing from the above, the use of eggshell waste will not only reduce the environmental problems associated with disposing eggshells in landfills, but will also more likely reduce the high cost associated with treating chronic wounds and burns.

10.7 Summary and Recommendations

The findings of this chapter have highlighted the different benefits of eggshell waste in the medical and dental fields, particularly its use to develop economical and value added products such as hydroxyapatite, collagen, and calcium supplements. From an environmental sustainability and management perspective, using eggshell waste material in the healing of segmented bone defects in orthopaedic and trauma surgery further strengthens the economic benefits associated with using natural waste material, which is high on the global agenda for a greener environment. Notably, this chapter showed that eggshells calcium is the best source of calcium and it is highly absorbable by the human body. Although eggshell waste is a convenient "at-home" calcium supplement, protocols such as washing and boiling the collected eggshells need to be correctly applied to kill common microbes such as *Salmonella* and *E. coli*. Considering the benefits of eggshell waste outlined in this chapter, it is highly recommended that more research be done on reducing the indiscriminate disposal of eggshells waste. Ultimately, this could create local employment and increase returns for poultry farmers.

References

1. Stadelman, W. J. (2000). Egg and egg products. In: Francis, F. J. ed. *Encyclopedia of Food Science and Technology.*, New York: John Wiley and Sons, pp. 593–599.
2. Cristianne, M. M., and Hincke, M. T. (2011). Recent patents on eggshells: Shell and membrane applications. *Recent Pat on Food Nutr Agric.*, 3, pp. 1–8.
3. Choi, H.-J., and Lee, S.-M. (2015). Heavy metal removal from acid mine drainage by calcined eggshells and microalgae hybrid system. *Environ. Sci. Pollut. Res.*, 22, pp. 13404–13411.

4. Bitzer, C. C., and Sims, J. T. (1998). Estimating the availability of nitrogen in poultry manure through laboratory and field studies. *J. Environ. Qual.*, 17, pp. 47–54.
5. Das, K., Minkara, M., Melear, N., and Tollner, E. (2002). Effect of poultry litter amendment on hatchery waste composting. *J. Appl. Poult. Res.*, 1, pp. 282–290.
6. Abdulrahman, I., Tijani, H. I., Mohammed, H. S., Yusuf, H., Jibrin, M. N., and Mohammed, S. (2014). From garbage to biomaterials: An overview on egg shell based hydroxyapatite. *J. Mater.*, 2014, pp. 1–6.
7. Oliveira, D. A., Benelli, P., and Amante, E. R. (2013). A literature review on adding value to solid residues: Egg shells. *J. Clean. Prod.*, 46, pp. 42–47.
8. Boron, L. (2004). Calcium citrate eggshells: Bioavailability and use as a dietary supplement. M.Sc, Federal University of Santa Catarina.
9. Murakami, F. S., Rodrigues, P. O., Campos, C. M. T., and Silva, M. A. S. (2007). Physicochemical study of $CaCO_3$ from eggshells. *Ciência e Tecnologia de Alimentos*, 27, pp. 658–662.
10. Fátima, A., Baptistella, L. H. B., Pilli, R. A., and Modolo, L. V. (2005). Ácidos siálicos–da compreensão do seu envolvimento em processos biológicos ao desenvolvimento de fármacos contra o agente etiológico da gripe. *Quim. Nova.*, 28, pp. 306–316.
11. Dutta, S., Passi, D., Singh, P., and Bhuibhar, A. (2015). Ceramic and non-ceramic hydroxyapatite as a bone graft material: A brief review. *Ir. J. Med. Sci.*, 184, pp. 101–106.
12. Giron, D. (2002). Applications of thermal analysis and coupled techniques in pharmaceutical Industry. *J. Therm. Anal. Calorim.*, 68, pp. 335–357.
13. Campos, C. M. T., Hamad, A. J. S., Block, J. M., Padilha, J. C. F., Ogliari, P. J., Barreira-Arellano, D., and Moreira, R. N. C. (2004). Composition of w-3 and w-6 fatty acids in freeze-dried chicken embryo eggs with different days of development. *Braz. Arch. Biol. Technol.*, 47, pp. 219–224.
14. Cardosa, T. M., Rodrigues, P. O., Stulzer, H. K., Segatto Silva, M. A., and Rosario Matos, J. (2005). Physical-chemical characterisation and polymorphism determination of two Nimodiphine samples deriving from distinct laboratories. *Drug Dev. Ind. Pharm.*, 31, pp. 631–637.
15. Holanda, J. N. F., and Freire, M. N. (2006). Characterisation of an avian eggshells waste aiming its use in a ceramic wall tile paste. *Ceremica*, 52, pp. 240–244.

16. Hincke, M. T., Nys, Y., Gautron, J., Rodriguez-Navarro, A. B., and Mckee, M. D. (2012). The eggshells: Structure, composition and mineralisation. *Front Biosci.*, 17, pp. 1266–1280.

17. Chien, Y.-C., Hincke, M., and McKee, M. (2008). Avian eggshells structure and osteopontin. *Cells Tissues Organs*, 189(1–4), pp. 38–43.

18. Kang, J., Kotaki, M., Okubayashi, S., and Sukigara, S. (2010). Fabrication of electrospun eggshells membrane nanofibers by treatment with catechin. *J. Appl. Polym. Sci.*, 117, pp. 2042–2049.

19. Dennis, J. E., Xiao, S. Q., Agarwal, M., Fink, D. J., Heuer, A. H., and Caplan, A. I. (1996). Microstructure of matrix and mineral components of eggshells from white leghorn chickens (Gallus gallus). *J. Morphol.*, 228, pp. 287–306.

20. Dorozhkin, S. V. (2013). Calcium orthophosphate-based bioceramics. *Materials.*, 6, pp. 3840–3942.

21. Lee, S.-W., Kim, S.-G., Balázsi, C., Chae, W.-S., and Lee, H.-O. (2012). Comparative study of hydroxyapatite from eggshells and synthetic hydroxyapatite for bone regeneration. *Oral surg. Oral Med. Oral Pathol.*, 113, pp. 348–355.

22. Prabakaran, K., Balamurugan, A., and Rajeswari, S. (2005). Development of calcium phosphate based apatite from hen's eggshells. *Bull. Mater. Sci.*, 28, pp. 115–119.

23. Ho, W.-F., Hsu, H.-C., Hsu, S.-K., Hung, C.-W., and Wu, S.-C. (2013). Calcium phosphate bioceramics synthesized from eggshells powders through a solid state reaction. *Ceram Int.*, 39, pp. 6467–6473.

24. Kamalanathan, P., Ramesh, S., Bang, L., Niakan, A., Tan, C., Purbolaksono, J., Chandran, H., and Teng, W. (2014). Synthesis and sintering of hydroxyapatite derived from eggshells as a calcium precursor. *Ceram Int.*, 40, pp. 16349–16359.

25. Akram, M., Ahmed, R., Shakir, I., Ibrahim, W. A. W., and Hussain, R. (2014). Extracting hydroxyapatite and its precursors from natural resources. *J. Mater. Sci.*, 49, pp. 1461–1475.

26. Wu, S.-C., Tsou, H.-K., Hsu, H.-C., Hsu, S.-K., Liou, S.-P., and Ho, W.-F. (2013). A hydrothermal synthesis of eggshells and fruit waste extract to produce nanosized hydroxyapatite. *Ceram Int.*, 39, pp. 8183–8188.

27. Ibrahim, A.-R., Li, X., Zhou, Y., Huang, Y., Chen, W., Wang, H., and Li, J. (2015). Synthesis of spongy-like mesoporous hydroxyapatite from raw waste eggshells for enhanced dissolution of ibuprofen loaded via supercritical CO_2. *Int. J. Mol. Sci.*, 16, pp. 7960–7975.

28. Rivera, E. M., Araiza, M., Brostow, W., Castano, V. M., Dýaz-Estrada, J., Hernandez, R., and Rodrýguez, J. R. (1999). Synthesis of hydroxyapatite from eggshells. *Mater. Lett.*, 41, pp. 128–134.

29. Zhang, C., Yang, J., Quan, Z., Yang, P., Li, C., Hou, Z., and Lin, J. (2009). Hydroxyapatite nano-and microcrystals with multiform morphologies: Controllable synthesis and luminescence properties. *Cryst. Growth Des*, 9, pp. 2725–2733.

30. Hui, P., Meena, S. L., Singh, G., and Satya Prakash, A. (2010). Synthesis of hydroxyapatite bio-ceramic powder by hydrothermal method. *J. Minerals Mater. Characterisation Eng.*, 9, pp. 683–692.

31. Khandelwal, H., and Prakash, S. (2016). Synthesis and characterization of hydroxyapatite powder by eggshells. *J. Minerals Mater. Characterization Eng.*, 4, pp. 119.

32. Kumar, G. S., Thamizhavel, A., and Girija, E. (2012). Microwave conversion of eggshells into flower-like hydroxyapatite nanostructure for biomedical applications. *Mater. Lett.*, 76, pp. 198–200.

33. Toriyama, M., Ravaglioli, A., Krajewski, A., Celotti, G., and Piancastelli, A. (1996). Synthesis of hydroxyapatite-based powders by mechano-chemical method and their sintering. *J. Eur. Ceramic Soc.*, 16, pp. 429–436.

34. Wu, S.-C., Hsu, H.-C., Hsu, S.-K., Chang, Y.-C., and Ho, W.-F. (2016). Synthesis of hydroxyapatite from eggshells powders through ball milling and heat treatment. *J. Asian Ceramic Soc.*, 4, pp. 85–90.

35. Ramesh, S., Natasha, A., Tan, C., Bang, L., Ching, C., and Chandran, H. (2016). Direct conversion of eggshells to hydroxyapatite ceramic by a sintering method. *Ceram Int.*, 42, pp. 7824–7829.

36. Kamalanathan, P., Ramesh, S., Bang, L., Niakan, A., Tan, C., Purbolaksono, J., Chandran, H., and Teng, W. (2014). Synthesis and sintering of hydroxyapatite derived from eggshells as a calcium precursor. *Ceram Int.*, 40, pp. 16349–16359.

37. Macha, I. J., Ozyegin, L., Oktar, F. N., and Ben-Nissan, B. (2015). Conversion of ostrich eggshells (struthio camelus) to calcium phosphates. *J. Aust. Ceram.*, 51, pp. 125–133.

38. Kattimani, V., Lingamaneni, K. P., Chakravarthi, P. S., Kumar, T. S., and Siddharthan, A. (2016). Eggshells-derived hydroxyapatite: A new era in bone regeneration. *J. Craniofac. Surg.*, 27, pp. 112–117.

39. Kim, S., Kim, W., Cho, J., Oh, N. S., Lee, M. H., and Lee, S. J. (2008). Comparison of bone formation in rabbits using hydroxyapatite and β-tricalcium phosphate scaffolds fabricated from egg shells.

In: *Proceedings of Advanced Materials Research. Trans Tech Publ*, pp. 999–1002.

40. Kattimani, V. S., Chakravarthi, P. S., Kanumuru, N. R., Subbarao, V. V., Sidharthan, A., Kumar, T. S., and Prasad, L. K., (2014). Eggshells derived hydroxyapatite as bone graft substitute in the healing of maxillary cystic bone defects: A preliminary report. *J Int. Oral Health.*, 6, p. 15.
41. Broadus, A. E. (1996). Mineral balance and homeostasis. In: Favus, M. J. ed. *Primer on the Metabloic Bone Diseases and Disorders of Mineral Metabolism*. 3rd edn. Philadelphia: Lippincott Raven, pp. 57–63.
42. Gaonkar, M., and Chakraborty, A. P. (2016). Application of eggshells as fertiliser and calcium supplement in tablets. *Int. J. Innovation Res. Sci. Eng. Technol.*, 5, pp. 3520–3525.
43. Brun, L. R., Lupo, M., Delorenzi, D. A., Di Loreto, V. E., and Rigalli, A. (2013). Chicken eggshells as suitable calcium source at home. *Int. J. Food Sci. Nutr.*, 64, pp. 740–743.
44. Hassan, N. M. (2015). Chicken eggshells powder as dietary calcium source in biscuits. *World J. Dairy Food Sci.*, 10, pp. 199–206.
45. Oguido, A. K., De Angelis, E. C., and Yada, I. F. U. (1995). Kinetics of intestinal calcium absorption from milk and eggshells in the rats. *Semina: Ciências Agrárias*, 16, pp. 7–13.
46. Ruff, K. J., Endres, J. R., Clewell, A. E., Szabo, J. R., and Schauss, A. G. (2012). Safety evaluation of a natural eggshells membrane-derived product. *Food Chem. Toxicol.*, 50, pp. 604–611.
47. Schaafsma, A., and Pakan, I. (1999). Effect of a chicken egg shell powder enriched dairy product on bone mineral density in persons with osteoporosis or osteopenia. *Nutrition-Burbank then Syracuse*, 15, pp. 157–157.
48. Hirasawa, T., Omi, N., and Ezawa, I. (2001). Effect of 1-hydroxyvitamin D3 and egg-shell calcium on bone metabolism in ovariectomized osteoporotic model rats. *J. Bone Miner. Metab.*, 2, pp. 84–88.
49. Association of American Feed Control Officials, (2009). Section 9.0 – Animal products, in: AAFCO Official Publication. Association of American Feed Control Officials Incorporated, pp. 341–348 (Chapter 5).
50. Vlad, V., and Biova, L. L. C. (2009). *Eggshells membrane separation method*. USA patent, 7534909.
51. Bee, W. (2012). How to make calcium from eggshells *Healing naturally by Bee* (Blog). Available: http://www.healingnaturallybybee.com (accessed September 3 2014).

52. King'Ori, A. M. (2011). A review of the uses of poultry eggshells and shell membranes. *Int. J. Poult. Sci.*, 10, pp. 908–912.

53. Suzuki, F., Wang, P. Y., Weatherspoon, J., and Mead, L. (2006). *Method of producing eggshells powder*. US patent, 20060062857 A1.

54. Makai, F., and Chudacek, J. (1991). The treatment of osteoporosis with Biomin-H. *Arch. Gerontol. Geriatr.*, pp. 487–490.

55. European Commision. 2015. *From by-product to pharmaceutical ingredient: Cracking eggshells waste* (online). The EU framework programme for research and innovation. Available: https://ec.europa.eu/programmes/horizon2020/en/news/product-pharmaceutical-ingredient-cracking-eggshells-waste (accessed 20 September 2016).

56. Rose-Martel, M., and Hincke, M. (2013). Eggshells as a source of novel bioactive molecules. *J. Food Sci. Eng.*, 3, p. 219.

57. Zhao, Y.-H., and Chi, Y.-J. (2009). Characterization of collagen from eggshells membrane. *Biotechnolo. J.*, 8, pp. 254–258.

58. Madison, N. (2011). *What is collagen?* 2003–2011 Conjecture Corporation.

59. Long, F. D., Adams, R. G., and DeVore, D. P. (2005). *Preparation of hyaluronic acid from eggshells membrane*: Google Patents.

60. Ponkham, W., Limroongreungrat, K., and Sangnark, A. (2011). Extraction of collagen from hen eggshells membrane by using organic acids. *Thai. J. Agr. Sci.*, 44, pp. 354–360.

61. Hussain, A. (2010). *Dielectric properties and microwave assisted separation of eggshells and membrane*. Library and Archives Canada, Bibliothèque et Archives Canada.

62. Nuutila, K., Katayama, S., Vuola, J., and Kankuri, E. (2014). Human wound-healing research: Issues and perspectives for studies using wide-scale analytic platforms. *Adv. Wound Care.*, 3, pp. 264–271.

Chapter 11

Production of Tamarind Products and Polysaccharide

Maw Maw Khaing, Nang Seng Moe, Kyaw Nyein Aye, and Nitar Nwe

Ecological Laboratory, Advancing Life and Regenerating Motherland, Yangon, Myanmar

mawmawkhaing116@gmail.com, nitarnwe@gmail.com

Tamarind tree grows in various parts of the world, especially in tropical regions. The parts of the tamarind tree have been used in many applications. The pulp of the tamarind fruit is an important component in chutneys, pickles, jams, curries, sauces, juices, instant powder, and seasoning for various dishes and is also used for washing hair for removing dandruff. The leaves and the flowers are eaten as vegetables and used to make curries, salads, and soups. The trunk of the tamarind tree is used for making furniture, wheels, axles, mill gears, handles, walking sticks, toy, oil presses, and sugar presses. Tamarind seed polysaccharide (TSP) obtained from tamarind seed is in the form of powder that can be used as an effective suspending agent. It is widely used in food additives as gelling material, glue for textile industries, and pharmaceuticals

Handbook of Biopolymers: Advances and Multifaceted Applications
Edited by Shakeel Ahmed, Suvardhan Kanchi, and Gopalakrishnan Kumar

ISBN 978-981-4800-17-4 (Hardcover), 978-0-429-02475-7 (eBook)
www.panstanford.com

formulation. TSP possesses properties such as high viscosity, broad pH tolerance, no carcinogenicity, adhesive nature, and biocompatibility and has two different forms for swelling and gelling properties. Thus, tamarind pulp and tamarind seed polysaccharides can be either produced as value-added foods or used for food additives.

11.1 Introduction

Tamarind, *Tamarindus indica* Linn is cultivated in most of the tropical regions. It is native to tropical Africa, where it grows wild throughout Sudan. It was so long ago introduced into India that it is also reported indigenous to India. Because of its many uses, its cultivation has spread around the world in tropical and subtropical zones. This evergreen tree is cultivated throughout India, Myanmar, Yemen, and Basra. At present, tamarind is cultivated in 54 countries of the world: 18 in its native range and 36 other countries where it has become naturalized. The major areas of production are Cameroon, Central African Republic, Ethiopia, Guinea, Kenya, Nigeria, Senegal, Sudan, Tanzania, Uganda, Afghanistan, Australia, Bangladesh, Brazil, Cambodia, China, Colombia, Cuba, Egypt, Malaysia, Mexico, Myanmar, Nicaragua, Pakistan, Philippines, Sri Lanka, India, and Thailand [1].

Tamarind is a large, slow-growing, long-living evergreen tree. The bark of the tamarind tree is brown-gray colored. It can tolerate diversity of soils such as loam, sandy, and clay soil, but well-drained slightly acidic soil is best for its growth [2]. It attains a height of 60–80 feet and bears a very large, widely spread head of foliage, trunk with a dark rough bark, and youngest twigs smooth or slightly pubescent; the flowers are in bunches, yellow in color, and boat shaped; the seeds are reddish brown and thick; the flattened sides of the seeds are marked by a centrally placed dull area; the fruit pulp occurs as a reddish-brown, moist, sticky mass, in which yellowish-brown fibers are readily seen; the odor is pleasant; the taste is sweetish and acidic; the bark of the trunk is scaly; the leaves are paripinnate up to 0.49 feet [3].

The tamarind tree produces fruits when the tree is 3–5 years old. Indehiscent fruits are pods curved and with protuberances due to the seed inside. They are 8–20 cm long. The pericarp is thin, tan or brown, brown-gray, which becomes a brittle shell at fruit maturity. Tamarind seeds are hard, brown, and wrapped in protective endocarp layer [4]. Tamarind fruit pulp contains up to 98% of tartaric acid and soluble sugar. It rich in pectin, organic acids, vitamin B, minerals, carotene, and vitamin C. Tamarind leaves are rich in minerals, such as potassium, phosphorus, calcium, and magnesium and a source of vitamin C and ß-carotene. Tamarind seeds are rich in protein and minerals, such as calcium, phosphorus, and magnesium [5]. This chapter describes the structure of tamarind tree, the production of tamarind pulp powder by using drum dryer, tamarind jam preparation process, production of tamarind paste, production of tamarind candy, preparation of tamarind kernel powder, extraction of tamarind seed polysaccharide (TSP), uses of tamarind seed polysaccharide, and products and application of the parts of tamarind tree.

11.2 The Structure of Tamarind Tree

The structure of tamarind tree comprises bark and trunk; leaves and flowers; and fruit and seeds. These have useful properties and different characteristics. Tamarind seed is a by-product of tamarind fruit and can also be used for making tamarind seed polysaccharide [1]. Leaves, flowers, and bark are useful for making mordants for dyeing and the trunk is used for construction purposes [2].

11.2.1 Leaves and Flowers of Tamarind Tree

The color, shape, and the way leaves grow differ from one type to another of the tree. By studying these one can quickly recognize each one. In hot countries, it is usual to see young new leaves reddish brown or pale pink in color. This may be a protection against strong sunlight, which could damage them. However, most full-grown leaves are green as they contain chlorophyll. This is the substance that helps the plant to convert carbon dioxide and water into sugar. The flowers of tamarind tree appear in the cold season, but many trees have a few flowers throughout the year. The flowers are five-petalled, lemon-yellow in color and set in a green cup. Inside the flowers are patches of deep maroon and purple. As the flowers fade, they turn from orange-pink to purple. Male and female flowers grow together on a branched stalk, which starts from the point where the leaves join the trunk. There are many of them and they are yellowish-orange and look like small Catkins (spike of soft, feathery flowers hanging from twigs of trees like willows and birches). The male flowers are smaller, there are more of them, and they have a scent [7].

11.2.2 Fruits and Seeds of Tamarind Tree

The tamarind fruit is a pod and has an outer epicarp, which is light gray or brown and scaly. The fruit pulp is firm but soft, thick, and blackish brown. The inside of the pulp contains the seeds [8]. Tamarind fruit pulp prevents the rancidity and to prolong the shelf life of coconut and groundnut oil [9]. Tamarind fruit is eaten fresh or processed and used as a seasoning or spice and in medicinal beverages [10]. Moreover, it has a sweet-sour pulp and the fibrous texture is used for manufacturing ice cream, soft drinks, and concentrated juices [11]. The fruit pulp is eaten fresh, suitable for fruit drink and garnish for fish [12]. Physico-chemical and mineral compositions of the whole seed and the seed nut of tamarind are shown in Table 11.1.

Tamarind indica seed contains a crude lipid 6.94–11.43% [13] and produces a large quantity of oil and fats for human consumption or industrial purposed [13]. Tamarind seeds consist of phenolic antioxidants, such as 2-hydroxy-30, 40-dihydroxyacetophenone, methyl 3,4-dihydroxybenzoate, 3,4-dihydroxyphenyl acetate and epicatechin [14]. They also contain proteins and have been reported as a source of food or food ingredients. The seed coat (20–30%) and the kernel or endosperm (70–75%) are included in the tamarind seed. The tamarind seed is used as a raw material in the production of tamarind seed kernel powder (TKP), polysaccharide (jellose), adhesive, and tannin [8]. The tamarind seed powder is used in coffee. The tamarind seed coat has polyphenolic compounds and antioxidant activity and is a rich source of polyphenols and tannins. The seed coat is extracted by using various solvents such as ethyl acetate, aqueous ethanol, methanol and aqueous acetone methanol and hexane [15]. Antioxidants attack free radicals, which contain one or more unpaired electrons and, have high chemical reactivity, and are harmful to the body. Thus, antioxidants are important for human health [16].

11.2.3 Bark and Trunk of Tamarind Tree

The grayish-brown bark of the tamarind tree is rough and pronouncedly scaly, having both vertical and horizontal fissures. The fissures are deepened and darken as the tree ages. The bark on the lower trunks of the oldest trees eventually falls away. At which point, the tree often develops an ornate and contorted trunk that is pleasing in appearance. These old trunks have various combinations of gray, brown, and tan colors. Trunk diameters on the oldest trees are up to 6 feet or even more [6].

Table 11.1 Physico-chemical and mineral composition of whole seed and seed nut of tamarind [17]

Parameter	Whole Seed	Seed Nut
Moisture %	11.75 ± 0.25	19.90 ± 0.1
Ash content %	7.35 ± 0.55	9.51 ± 0.01
Carbohydrates %	17.10 ± 0.1	8.9 ± 0.9
Crude fiber %	3.82 ± 0.005	2.33 ± 0.02
Crude protein %	22.20 ± 0.55	21.25 ± 0.01
Sodium (Na)	2.1 ± 0.05	3.8 ± 0.05
Potassium (K)	21.0 ± 0.02	41.0 ± 0.01
Calcium (Ca)	10.0 ± 0.05	31.0 ± 0.02
Magnesium (Mg)	15.0 ± 0.05	13.2 ± 0.01

11.3 Extraction of Tamarind Pulp from Tamarind Fruits

In cold extraction, the tamarind flesh is soaked in water in a tamarind flesh:water ratio of 1:2 for 6 h. In hot extraction, the flesh is soaked in water at a temperature of 70°C for 10 min and then it is soaked for 6 hr. Then the homogenization and filtration of the soaked flesh is carried out to obtain fine pulp. The extracted pulp gives highest total soluble solids and total sugars. In these two methods, the extraction was maintained as a flesh:water rate of

1:2 [18]. The effect of soaking media and period on pulp recovery, total soluble solid (TSS), ascorbic acid, and total sugars of tamarind pulp is shown in Table 11.2.

Table 11.2 Effect of soaking media and period on pulp recovery, TSS, ascorbic acid and total sugars of tamarind pulp [19]

	Soaking media	Pulp recovery (%)	TSS (%)	Ascorbic acid (%)	Total sugar (%)
1	Soaking in cold water (1.5 h)	51.61	14.73	5.69	19.73
2	Soaking in cold water (3 h)	57.7	15.71	5.31	20.83
3	Soaking in hot water (1.5 h)	73.46	19.83	4.83	22.67
4	Soaking in hot water (3 h)	81.26	22.18	4.78	23.52

11.4 Using Tamarind Pulp

The extracted tamarind pulp contains flavonoids, tanins and alkaloids, steroids, saponin, glycosides, proteins, amino acids and reducing sugar were not contain [20]. The chemical composition of tamarind pulp is shown in Table 11.3. Tamarind fruit pulp (*Tamarindus indica* L.) is used for a wide variety of domestic and industrial purposes [21]. Tamarind pulp can be used as a raw material for the production of tamarind juice concentrate, tamarind pulp powder, tartaric acid, and pectin [22].

It is used for food additives such as seasoning, to flavor confections, curries, and sauces, and is a main component in juices and certain beverages [21].

Table 11.3 The chemical composition of tamarind pulp [21]

No	Characteristics	Composition (%)
1	Water	20.6
2	Protein	3.1
3	Fat	0.4
4	Fiber	3
5	Ash	2.1

11.5 The Production of Tamarind Pulp Powder by Using Drum Dryer

The sour tamarind flesh is mixed with distilled water and then the mixture is crushed and the juice is separated by the pulper finisher. The concentration of fresh juice is adjusted to 20° Brix and the color and pH are measured. The obtained tamarind juice is warmed. During warming, after the addition of the drying aid (Maltodextrin, MD; and Arabic Gum, AG), the mixture is stirred and dried at 120 and 140°C. The ratio of the drying aid and the tamarind juice (20°Brix) is 1:0.4, 1:0.8, and 1:1.4. After warming, the mixture of the juice is fed into the drum dryer (drum speed of 0.35 rpm and the gap between drums of 0.4 mm). After that, the tamarind pulp powder is obtained and bulk density, moisture content, water activity and solubility are determined. The tamarind pulp powder is reconstituted by using distilled water, and the pH, color, percentage of insoluble solid, and viscosity are also determined [23].

11.6 Tamarind Jam Preparation Process

The fresh tamarind fruit is washed and peeled. The peeled tamarind is used to extract the pulp. The extracted pulp is blended with 20% water by using a blender. The blended pulp is boiled for 30 min to evaporate the water. Sugar is added gradually until the total soluble solid reaches 68%. Then the tamarind jam is obtained and cooled at 80°C before bottling. Finally, the cooled tamarind jam is filled into bottles and the bottles are sealed tightly [24].

11.7 Production of Tamarind Paste

The tamarind pulp is soaked in water for 4 h. The soaked pulp is repeatedly squeezed until the juice is extracted and passed through a 1 mm sieve. One hundred grams of extracted juice is boiled to obtain the totally soluble solid at 32–34° Brix. At the end of boiling, ingredients such as 20 g salt, 10 ml + 0.02 g of butylated hydroxyl anisole are added and mixed thoroughly. The final product 200 g of paste is filled hot into the pre-sterilized pet jar and sealed air tight [25].

11.8 Production of Tamarind Candy

First, fresh tamarind fruits are cleaned and the shells are removed manually by breaking them open with hands. The tamarind seeds are removed and the pulp is subjected to size reduction to obtain tamarind pulp power. Ingredients such as 150 g of sugar, 20 g of glucose syrup, 20 g of tamarind pulp powder, and 50 ml of water are used. After that, these ingredients are combined in a saucepan over medium heat. The sugar is stirred until it is dissolved. The mixture of the ingredients is boiled without stirring until the temperature of the candy reaches 105–110°C for 1 hr. When the mixture of the temperature reaches 110°C, the candy is removed and allowed to sit until it stops bubbling completely. The candy is poured into greased mould cavities and is then allowed to cool completely and removed when hardened. Finally, the candies are packed in a foil and stored in an airtight container at room temperature [26].

11.9 Preparation of Tamarind Kernel Powder

Normally tamarind seeds can be collected from tamarind products production area. Firstly, seeds were dried in hot air oven at 60°C for 6 hr and than roasted at about 120°C for 15 min. After the roasted seeds were cooled, shells of seeds were cracked out using an electrical stone motor. The remaining tamarind seed kernels, 1 kg were soaked in 3 L of water for about 6 hr. After that tamarind seed kernels were washed and drained. The soaked seed kernels were grounded to a smaller particle size using a blender coarsely. The grounded kernels were dried in hot air oven at 50–55°C for 6 hr and then grounded in a pin mill through a 0.25 mm sieve size to obtain fine powder [27]. The resultant tamarind seed kernel powder were used for extraction of tamarind seed polysaccharide.

11.10 Extraction of Tamarind Seed Polysaccharide

Tamarind kernel powder (20 g) is added to 200 ml of cold distilled water to prepare a slurry to isolate the tamarind seed polysaccharide and then 800 ml of boiled distilled water is added to the slurry. The solution is boiled for 20 min with continuous stirring. The resulting solution is kept overnight and then centrifuged at 5000 rpm for 20 min. The mixture of the supernatant liquid and the precipitate is obtained. The supernatant liquid is poured into twice the volume of absolute alcohol with continuous stirring. The obtained precipitate is washed with absolute ethanol and air-dried. The dried polymer is powdered, passed through sieve no. 60 and stored in a desiccator until for further use [28].

11.11 Uses of Tamarind Seed Polysaccharide

Natural polymers are chemically inert, nontoxic, less expensive, biodegradable, and widely available and have less chance of adverse effects [29]. Polymers are macromolecules composed of a large number of repeating monomer units and can be used in different pharmaceutical dosage forms as diluents, binders, disintegrants, suspending agents, sustained release agents, stabilizers, and coating materials.

Tamarind seed polysaccharide contains 50–60% of the polysaccharide in gel form in the presence of sugar or alcohol [30]. The polysaccharides are monomers of glucose, galactose and xylose in molar ratio of 3:1:5 [31, 32]. The polysaccharides obtained from tamarind seed kernel have the characteristic property of forming gels with sugar concentrates, such as fruit pectin, and mucilaginous dispersion with water. Tamarind polysaccharides are not regarded as true pectin because they do not contain galacturonic acid and methyluonate [33].

Tamarind seed polysaccharides are complex carbohydrates. Carbohydrates provide good mechanical properties for use as fibers, films, adhesives, hydrogels, emulsifiers, and drug delivery agents and can be used to form pectin like gels, jams, and jellies [34]. TSP improves dry eye syndrome and assists in releasing drugs in the human body and intraocular penetration of Rufloxacin [35]. The purified polysaccharide is free from methyl ester groups and galacturonic acid units and is designated as a "polyose" or "hexo-pentosan," which may be specifically termed gluco-galacto-xylan [36]. Tamarind seed polysaccharides are isolated with a galactoxyloglucan, which is contained in the seed kernel of *Tamarindus indica.* It possesses high viscosity, broad pH tolerance, and adhesive properties. This can be used in food and pharmaceutical industries as stabilizer, thickener, gelling agent, and binder [37].

Tamarind seed polysaccharides are applied in personal care products as moisturizing ingredients that can also improve the skin feel of the product. The study of toxicological indicates that xyloglucan is very well tolerated by conjunctival cells and can reduce drug-related toxicity. Moreover, xyloglucan promotes corneal wound healing due to its greater interaction with the integrin recognition system [38].

The organoleptic properties of tamarind seed polysaccharide are color, odor, taste, touch, fracture, and texture [39].

Tamarind seed polysaccharide consists of a cellulose-like backbone that carries xylose and galactoxylose substitutes and it has been identified as a non-ionic, neutral, and branched polysaccharide [40]. Tamarind seed polysaccharide gives the product a mucin-like molecular structure and mucoadhesive properties [40]. About 80% of glucose residues are substituted by xylose residues (1–6 linked), which themselves are partially substituted by p-1-2 galactose residues (Fig. 11.1). The exact sequential distribution of branches is not known. TSP is a branched polysaccharide with a main chain of Â-D-1-glucopyrynosyl units, with a side chain consisting of single D-xylopyranosyl unit attached to every second, third, and fourth D glucopyrynosyl unit through the 1-6 linkage as shown in Fig. 11.1 [41]. The comparison of the physico-chemical characteristics of raw tamarind seed and extracted tamarind seed polysaccharides is shown in Table 11.4. The products and application of the parts of tamarind tree are shown in Table 11.5.

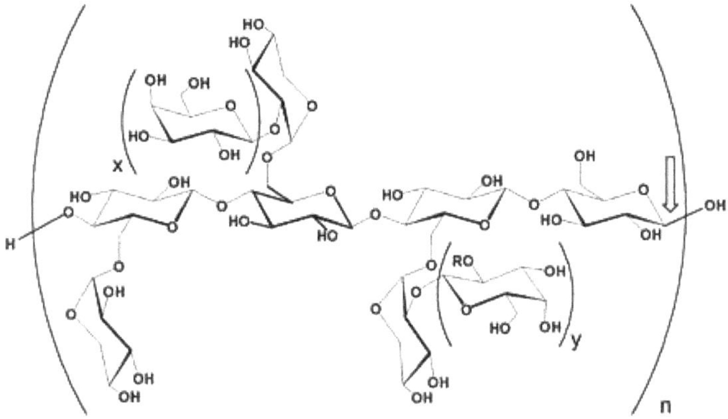

Figure 11.1 Chemical structure of tamarind seed polysaccharide.

Table 11.4 Compare the physico-chemical characteristics of raw tamarind seed and extracted tamarind seed polysaccharides [42]

	Raw tamarind seed		Tamarind seed polysaccharide	
Parameter (%)	**Light brown (LB)**	**Dark brown (DB)**	**Light brown (LB)**	**Dark brown (DB)**
Moisture	10.99b ± 0.1002	11.21a ± 0.0058	3.810 ± 0.070	3.800 ± 0.000
Protein	20.23b ± 0.5658	23.75a ± 0.0839	3.900 ± 0.000	6.500 ± 0.017
Oil	3.90a ± 0.0500	3.17b ± 0.0462	1.323 ± 0.070	1.360 ± 0.540
Ash	2.50a ± 0.0000	2.17b ± 0.0577	1.043 ± 0.060	2.933 ± 0.060
Fiber	3.08a ± 0.0493	2.51b ± 0.0058	1.007 ± 0.010	1.030 ± 0.660
Carbohydrate	59.303b ± 0.4041	57.33a ± 0.2452	88.850 ± 0.600	85.210 ± 0.530

Note: Each value is a mean of three determinations.

Table 11.5 Products and application of the parts of tamarind tree

Parts of tamarind tree	Products	Application	Refs
Seed	Tamarind kernel powder	Textile, paper sizing, leather treating, glue for wood	[43]
Flower and Leave	Mordants	Dyeing	[43]
Seed Coat	Ethyl Acetate	Antioxidant in herbals	[44]

(*Continued*)

Table 11.5 (*Continued*)

Parts of tamarind tree	Products	Application	Refs
Fruit pulp	Pulp powder	Beverages, curries and sauces	[45]
Trunk	wood	General used in construction and to make products such as boats, tool handles, furniture, and turnery	[6]

11.12 Conclusion

In this chapter, the structures of the tamarind tree, such as bark and trunk, leaves and flowers, and fruit and seeds, were studied. A variety of tamarind products have been produced using whole parts of the tamarind tree. These products include tamarind pulp powder, tamarind paste, tamarind jam, and tamarind candy. Tamarind seed polysaccharide can be obtained from tamarind seed kernel by drying and grinding. Tamarind seed polysaccharide has been used in many applications and researchers are now paying more attention to the study of the applications of tamarind seed polysaccharide.

References

1. Singh, D., Wangchu, L., and Moond, S. K. (2007). Processed Products of Tamarind, *Natural Product Radiance,* 6(4), pp. 315–321.
2. Isha, D., and Milind, P. (2012). Imlii: A Craze Lovely, *International Research Journal of Pharmacy,* 3(8), pp. 110–115.
3. Tariq, M., Chaudhary, S. S., Rahman, K., Hamiduddin, Zaman, R., and Imtiyaz, S. (2013). *Journal of Biological & Scientific Opinion,* 1(2), pp. 128–131.
4. Karmegam, N., Kalpana. M., and Prakash. M. (2014). Allelopathic Effect of Aqueous Root Bark Extract of *Tamarindus indica* L. and Rhizosphere Soil on Germination and Seedling Growth of *Oryza sativa* L., *International Journal of Current Microbiology and Applied Sciences,* 3(3), pp. 505–514.

5. Stege, C. V. D., Prehsler, S., Hartl, A., and Vogl, C. R. (2011). Tamarind (*Tamarindus indica* L.) in the Traditional West African Diet: Not Just a Famine Food, *Fruits,* 66, pp. 171–185.
6. Brown, S. H. (2013), *Tamarind indica*, Florida, *IFAS Extension,* 239, pp. 533–7513.
7. Mukherjee, P. (1983). *Common Trees of India*, WWF Nature Guides, World Wildlife Fund, India.
8. Sahoo, S., Sahoo, R., and Nayak, P. L. (2010). Tamarind Seed Polysaccharide: A Versatile Biopolymer for Mucoadhesive Applications, *Journal of Pharmaceutical and Biomedical Sciences,* 8(20), pp. 1–12.
9. Vadivel, V., and Pugalenthi, M. (2010). Evaluation of Nutritional Value and Protein Quality of an Under-Utilized Tribal Food Legume, *Indian Journal of Traditional Knowledge,* 9(4), pp. 791–797.
10. Escalona-Arranz, J. C., Pérez-Rosés, R., Jiménez, I. L., Rodríguez-Amado, J., Argota-Coello, H., Cañizares-Lay, J., Morris-Quevedo, H. J., and Sierra-González, G. (2010). Chemical Constituents of *Tamarindus indica* L. Leaves, *Revista Cubana De,* XXII(3), pp. 65–71.
11. Luzia, D. M. M., and Jorge, N. (2011). Antioxidant Activity, Fatty Acid Profile and Tocopherols of *Tamarindus indica* L. Seeds, *Ciência e Tecnologia de Alimentos-Campinas,* 31(2), pp. 497–501.
12. Ajayi, S. A., Odofin, W. T., Denton, L., Ogunbodede, B. O., Ado, S. G., Morakinyo, J. A., and Ladipo, D. O., Dumet, D., Ndirpaya, Y. N., Adetula, O., Aladele, S. E., Umar, M. L., and Zaria, A. A. (1996–2008). State of Plant Genetic Resources for Food and Agriculture in Nigeria.
13. Panchal, B., Deshmukh, S., and Sharma, M. (2014). Optimization of Oil Extraction and Characterization from *Tamarindus indica* Linn Seed Oil, *International Journal of Oil, Gas and Coal Engineering,* 2(1), pp. 1–6.
14. Thongmuang, P., and Sudjaroen, Y. (2013). The Tyrosinase and Cyclooxygenase Inhibitory Activities and Cytotoxicity Screening of *Tamarindus indica* Seeds, *International Journal of Biological, Biomolecular, Agricultural, Food and Biotechnological Engineering,* 7(1), pp. 11–13.
15. Nakchat, O., Meksuruyen, D., and Pongsamart, S. (2014). Antioxidant and Anti- Lipid Peroxidation Activities of Tamarind Indica Seed Coat in Human Fibroblast Cells, *India Journal of Experimental Biology*, 52, pp. 125–132.

16. Reis, P. M. C. L., Hense, H., Dariva, C., Franceschi, E., and Ângela, G., Vieira, B. (2013). Obtaining Antioxidant Compounds Seed *Tamarindus indica*, Sweet Variety, *Iberoamerican Conference on Supercritical Fluids Cartagena de Indias (Colombia)*, 3, pp. 1–7.

17. Yusu, A. A., Mofio, B. M., and Ahmed, A. B. (2007). Proximate and Mineral Composition of *Tamarindus indica* Linn 1753 Seeds, *Science World Journal*, 2(1), pp. 1–4.

18. Joshi, A. A., Kshirsagar, R. B., and Sawate, A. R. (2012). Studies on Standardization of Enzyme Concentration and Process for Extraction of Tamarind Pulp, Variety Ajanta, *Journal of Food Processing and Technology*, 3(2), pp. 1–3.

19. Karnataka (2013). Standardization of Methods for Extraction of Tamarind Pulp, *Karnataka Journal of Agricultural Sciences*, 26(4), pp. 570–571.

20. Paul Das, A. M., and Banerjee, A. (2014). Extraction of Tamarind Pulp and Its Antibacterial Activity, *Asian Journal of Plant Science and Research*, 4(2), pp. 47–49.

21. Caluwé, E. D., Halamová, K., and Damme, P. V. (2010). *Tamarindus indica* L.: A Review of Traditional Uses, Phytochemistry and Pharmacology, *Afrikafocus*, 23(1), pp. 53–83.

22. Ramchander, T., Rajkumar, D., Sravanprasad, M., Goli, V., and Arjun, C. H. D. (2012). Antidiabetic Activity of Aqueous Methanolic Extracts of Leaf of *Tamarindus indica*, *International Journal of Pharmacognosy and Phytochemical Research*, 4(1), pp. 5–7.

23. Jittanit, W., Chantara-In, M., Deying, T., and Ratanavong, W. (2011). Production of Tamarind Powder by Drum Dryer Using Maltodextrin and Arabic Gum as Adjuncts, *Songklanakarin Journal of Science and Technology*, 33(1), pp. 33–41.

24. Jimoh, S. O., and Onabanjo, O. O. (2012). Potentials of *Tamarindus indica* (Linn) in Jam Production, *Journal of Agriculture and Social Research (JASR)*, 2 (12), pp. 29–43.

25. Patter, A., Kukanoor, L., Hegde, N. K., and Jholgiker, P. (2013). Studies on Keeping Quality of Tamarind Paste during Storage, *The Asian Journal of Horticulture*, 8(2), pp. 430–432.

26. Obasi, N. E., Okorocha, C., and Orisakwe, O. F. (2013). Production and Evaluation of Velvet Tamarind (*Dialium Guineese Wild)* Candy, *European Journal of Food Science and Technology*, 1(1), pp. 1–8.

27. Sukhawanli, S., and Thamakorn, P. (2014). Extraction of Tamarind Seed Jellose under Different Conditions and Their Rheological Properties, *Food and Applied Bioscience Journal,* 2(1), pp. 61–68.

28. Deveswaran, R., Bharath, S., Furtado, S., Abraham, S., Basavaraj, B. V., and Madhavan, V. (2010). Isolation and Evaluation of Tamarind Seed Polysaccharide as a Natural Suspending Agent, *International Journal of Pharmaceutical & Biological Archives,* 1(4), pp. 360–363.

29. Razavi, M., Nyamathulla, S., Karimian, H., Moghadamtousi, S. Z., and Noordin, M. I. (2014). Hydrogel Polysaccharides of Tamarind and Xanthan to Formulate Hydrodynamically Balanced Matrix Tablets of Famotidine, *Molecules,* 19, pp. 13909–13931.

30. Marathe, R. M., Annapure, U. S., Singhal, R. S., and Kulkarni, P. R. (2002). Gelling Behaviour of Polyose from Tamarind Kernel Polysaccharide, *Food Hydrocolloids,* 16, pp. 423–426.

31. Jangdey, M. S., Gupta, A., Kaur, C. D., and Saraf, S. (2016). Assessment of Utilization, Value Addition and Characterization of Tamarind: A Natural Gum of Chhattisgarh, *International Journal of Pharmaceutical Research & Allied Sciences,* 5(2), pp. 323–334.

32. Katiyar, N., Malviya, R., and Sharma, P. K.(2015). Effect of Heating Conditions on the Physical Properties of Tamarind Seed Polysaccharide, *Global Journal of Pharmacology,* 9(1), pp. 47–50.

33. Chandra, S., and Samsher (2006). Medico-Nutritionally Enriched Fruit-Tamarind, *International Journal of Agriculture Sciences,* 2(2), pp. 635–638.

34. Chandramouli, Y., Firoz, S., Vikram, A., Mahitha, B., Yasmeen, B. R., and Hemanthpavankumar, K. (2012). Tamarind Seed Polysaccharide (TSP)-An Adaptable Excipient for Novel Drug Delivery System, *International Journal of Pharmacy Practice & Drug Research,* 2(2), pp. 57–63.

35. Mohamad, M. Y., Akram, H. B., and Bero, D. N. (2012). Tamarind Seed Extract Enhances Epidermal Wound Healing, *International Journal of Biology,* 4(1), pp. 81–88.

36. Savur, G. R., and Sreenivasan, A. (1947). Isolation and Characterization of Seed Tamarind (*Tamarindus indica* L.) Polysaccharide, *The Journal of Biological Chemistry,* 172(2), pp. 501–509.

37. Sumathi, S., and Ray, A. R. (2002). Release Behaviour of Drugs from Tamarind Seed Polysaccharide Tablets, *Journal of Pharmacy & Pharmaceutical Sciences,* 5(1), pp. 12–18.

38. Semenzato, A., Costantiniand, A., and Baratto, G. (2015). Green Polymers in Personal Care Products: Rheological Properties of Tamarind Seed Polysaccharide, *Cosmetics*, 2, pp. 1–10.

39. Bansal, J., Kumar, N., Malviya, R., and Sharma, P. K. (2014). Extraction and Evaluation of Tamarind Seed Polysaccharide as Pharmaceutical in situ Gel Forming System, *American-Eurasian Journal of Scientific Research,* 9(1), pp. 1–5.

40. Sahoo, R. (2015). Natural Polymer of Tamarind Seed: A Prospective Carrier for Ocular Drug Delivery, *European Scientific Journal,* 11(27), pp. 267–275.

41. Manchanda, R., Arora, S. C., and Manchanda, R. (2014). Tamarind Seed Polysaccharide and Its Modifications-Versatile Pharmaceutical Excipients: A Review, *International Journal of PharmTech Research,* 6(2), pp. 412–420.

42. Mohamed, H. A., Mohamed, B. E., and Ahmed, K. E. (2015). Physicochemical Properties of Tamarind (Tamarind indica) Seed Polysaccharides, *Food Processing & Technology,* 6(6), pp. 1–5.

43. Morton, J. F. (1958). The Tamarind (*Tamarindus indica* L.) Its Food, Medicinal and Industrial Uses, Florida State Horticultural Society.

44. Khairunnuur, F. A., Zulkhairi, A., Azrina, A., Moklas, M. A. M., Khairullizam, S., Zamree, M. S., and Shahidan, M. A. (2009). Nutritional Composition, in vitro Antioxidant Activity and *Artemiasalina* L. Lethality of Pulp and Seed of *Tamarindus indica* L. Extracts, *Malaysian Journal of Nutrition,* 15(1), pp. 65–75.

45. Pugalenthi, M., Vadivell, V., Gurumoorthi, P., and Janardhanan, K. (2004). Comparative Nutritional Evaluation of Little Known Legumes, *Tamarindus indica, Erythrina indica* and *Sesbania bispinosa, Tropical and Subtropical Agroecosystems,* (4), pp. 107–123.

Index